COLUMBIA NORTHWEST CLASSICS

Chris Friday, Editor

COLUMBIA NORTHWEST CLASSICS

Columbia Northwest Classics are reprints
of important studies about the peoples and places
that make up the Pacific Northwest. The series focuses
especially on that vast area drained by the Columbia River
and its tributaries. Like the plants, animals, and people
that have crossed over the watersheds to the east, west,
south, and north, Columbia Northwest Classics embrace
a Pacific Northwest that includes not only Oregon,
Washington, and Idaho but also British Columbia,
the Yukon, Alaska, and portions of Montana,
California, Nevada, and Utah.

Mountain Fever:
Historic Conquests of Rainier
by Aubrey Haines

To Fish in Common:
The Ethnohistory of Lummi Indian Salmon Fishing
by Daniel L. Boxberger

Mexican Labor and World War II:
Braceros in the Pacific Northwest, 1942–1947
by Erasmo Gamboa

Mountain Fever

HISTORIC CONQUESTS OF RAINIER

AUBREY L. HAINES

Foreword by Ruth Kirk

UNIVERSITY OF WASHINGTON PRESS

Seattle and London

TO PHILEMON BEECHER VAN TRUMP

"Van Trump has done more than all others combined
to interest his countrymen in the mountain."

OLIN D. WHEELER, *Wonderland—1895*

Copyright © 1962 by the Oregon Historical Society
Reprinted as a Columbia Northwest Classic in 1999
Foreword to the Columbia Northwest Classic edition copyright
© 1999 by the University of Washington Press
Printed in the United States of America

Library of Congress Cataloging-in-Publication Data
Haines, Aubrey L.
Mountain fever : historic conquests of Rainier /
Aubrey L. Haines ; foreword by Ruth Kirk.
p. cm. — (Columbia Northwest Classics)
Originally published: Portland : Oregon Historical Society, 1962.
Includes bibliographical references and index.
ISBN 0-295-97847-3 (acid-free paper)
1. Rainier, Mount (Wash.)—History—19th century.
2. Mount Rainier National Park (Wash.)—History.
3. Mountineering—Washington (State)—Rainier, Mount—History—19th century.
I. Title. II. Series: Columbia northwest classics (Seattle, Wash.)
F897.R2H3   1999   99-37124
979.7'782—dc21   CIP

The paper used in this publication is acid-free and recycled from
10 percent post-consumer and at least 50 percent pre-consumer waste.
It meets the minimum requirements of American National Standard
for Information Sciences—Permanence of Paper for Printed
Library Materials, ANSI Z39.48-1984. ♻ ⊗

CONTENTS

v

FOREWORD

While serving as a ranger at Mount Rainier National Park in the 1950s, Aubrey Haines began *Mountain Fever*. By the time he finished in the 1960s, he and his family were living in Yellowstone, where he served first as park engineer, then as park historian—showing a remarkable versatility.

We, too, lived at Rainier in the 1950s. My husband also was a ranger there, and I well remember Aubrey driving out the Nisqually gate on his days off, heading for the Washington State Historical Society in Tacoma, the State Library in Olympia, the Oregon Historical Society in Portland, or some other repository. He researched journals, letters, and historic newspapers and also interviewed old-timers. At the time, the story of the park's immediate prelude had not been drawn together for the public.

Early in 1999, I visited Aubrey and his wife, Wilma, now living in Tucson, and talked to him about the book. "People get all steamed up about climbing Rainier," Aubrey commented. "'Mountain fever' was a disease pioneers got just by crossing the plains." Thus, William Fraser Tolmie, in 1833 newly arrived at a Hudson's Bay Company fur-trade post near today's Tacoma, became the first of the newcomers to ask natives for guide service to the big mountain on the horizon. And thus, to celebrate the 1962 original publication of *Mountain Fever,* Thomas Vaughn, at the time director of the Oregon Historical Society, suggested an autograph party at the summit of Mount Rainier. Aubrey vetoed the prospect: he "had no wish to go again to the top, one plodding step after the other, with parched lips and a little nagging headache because my gasping lungs and trip-hammering heart couldn't keep up the supply of oxygen." Instead of a party at the summit, the destination would be Camp Muir, at 10,000 feet the high camp for summit climbers and a pleasant day-hike above road's end at Paradise. Alas, it was not to be. Looking out his Tucson window Aubrey reminisced: "There was a terrible doggone blizzard up there and partway to Muir I called every-thing off. That was a shame. We were carrying a huge baked salmon fitted onto a packboard and we had to just hunker in among some rocks and eat it fast, then hike back down."

The book the blizzard prevented from being signed on Rainier's high slopes is a scholarly product exactingly detailed and documented. It also brims with the action and enthusiasm of those earlier generations taken by mountain fever: it is lively reading. Although smitten, Haines was not blinded by the fever. In some ways he was ahead of his time as a writer. His account of the Northern Pacific and Tacoma Eastern railways' relation to the mountain is a chronicle of ecotourism's beginnings.[1] His discussion of the politics behind passage of park legislation is unromanticized. Regional political and corporate backing was crucial—and "corporate backing" can be read Northern Pacific. The company urged that all land within national parks should be federal, a concept that would "force" them to trade their land-grant sections at Rainier for acreage elsewhere. They chose Oregon's lush Douglas-fir forests—rock and ice in exchange for timber, in the words of one Northern Pacific official.

In 1893 Mount Rainier was designated as a forest reserve, a formal recognition of its scenic values and potential for tour-ism, but without protection from timber cutting, mining, or graz-ing.[2] Washington's congressional delegation made five tries to change the reserve status to a national park, which would pro-hibit commercial utilization of resources. In 1899 they suc-ceeded, but only after a backroom negotiation that resulted in a promise not to ask for operating funds. For four years no money was forthcoming: a national park, but no funds for managing it—a scenario that will sound familiar to today's park adminis-trators, who also have to cope with tight finances.

History involves perspective, and the telling of the tale changes with time. In one way, Haines' book, reprinted here without alteration, shows that it was written half a century ago. His rather one-dimensional depiction of Native Americans' relation to the mountain is not what most would pen today. The five language groups living within Rainier's vast radiating drainages regarded the great peak with reverence and regularly and gratefully traveled there to hunt, pick huckleberries, and gather material for basketry. These special feelings have not disappeared on either side of the mountain. A 1999 article in the *Yakima Herald-Republic* reveals that tribespeople to the east still feature Mount Rainier in songs and dances regarded as sacred. To the west of the peak, the Nisqually Tribe signed a special use agreement with the National Park Service in 1998. It secures free Nisqually access to the park to collect plants for food or medicine, or to use in traditional crafts or rituals.

As Haines describes, when newcomer fur traders, U.S. Army officers, travelers, land surveyors, and settlers asked for help in reaching the mountain, Indian men readily gave it. They knew the mountain well, as they knew all of their homeland and the trade routes that led to others' domains. Indian guides refused, however, to climb much above Rainier's subalpine meadows; to venture into the realm of perpetual snow and ice without great ritual preparation amounted to sacrilege. Their clients' brash persistence in doing so must have struck them as appallingly disrespectful and quite certain to bring disaster.

Indian and white interpretations of the treaties signed in 1854 and 1855 differed significantly.[3] Technically, all land was ceded to the federal government, which then reserved some for tribal use, allowed some to be claimed or bought by settlers, and set aside some for specific federal purposes. Mount Rainier belonged in the last category, but to native people its slopes must have looked as they had for generations. In the epilogue to *Mountain Fever*, Haines describes a 1915 instance of park rangers treating encamped Yakama Indians as trespassers: "Here was the noble redman, deprived of the rights guaranteed him by treaty and hounded from his ancestral hunting grounds—a poor recompense for loyally guiding the first white men to scale the mountain" (p. 205). Haines recognized the irony in the circumstances, but this crafting of the incident would quite surely be different were it written today—and that is heartening. In the last half-century we have progressed in crosscultural understanding, and the dated tone of this particular passage is in itself part of the broader historical tale.

We also have gained new understanding of Rainier as a volcano.[4] Native people recognized its distinctive force in their stories of the mountain as a wife so angry with her husband that she left him, moved off by herself, grew huge, and threw rocks spitefully at other peaks. Today's geologists speak of eruptions, which have come more often than previously realized. They also speak of snow that has combined with volcanic heat to form a lake within summit melt caverns, and of the collapse of the top 2,000 feet of the mountain that created a mudflow so massive it pushed the shoreline of Puget Sound seaward. Archaeologists corroborate this mudflow and date it at about 5,700 years ago. They have found artifacts beneath the mud. People were present.

In the years since that devastation, the human population ringing Mount Rainier has increased astronomically and the hazard has not decreased. Indeed, geologists now trace several mudflows that also have reached Puget Sound. Sensors placed near the mountain recently are intended to give warning of virtually certain, but unpredictable, volcanic activity or mudflows, and National Park Service officials debate how to balance public safety and ecological protection of the park with public love for the mountain and insistence on continued access, at will, by automobile. "Mountain fever" seems not to diminish. Haines' book reminds us how much and how little we have changed in the past century. What will our successors be celebrating when the park reaches its bicentennial?

Ruth Kirk

April 1999

1. Hal K. Rothman, *Devil's Bargains: Tourism in the Twentieth Century American West* (Lawrence: University Press of Kansas, 1998), provides a useful overview of ecotourism.

2. Richard West Sellars, *Preserving Nature in the National Parks* (New Haven: Yale University Press, 1997), discusses the roots of national park concepts, legislation, and evolving management from the late 1800s to the 1990s. Although national parks and national forests serve different purposes, shifting practices and policies in both reflect common scientific and public understandings in a given historical era. By way of comparison, also see Paul W. Hirt, *A Conspiracy of Optimism: Management of the National Forests since World War Two* (Lincoln: University of Nebraska Press, 1994); Dale Goble and Paul Hirt, eds., *Northwest Lands, Northwest Peoples: Readings in Environmental History* (Seattle: University of Washington Press, 1999); and Nancy Langston, *Forest Dreams, Forest Nightmares: The Paradox of Old Growth in the Inland West* (Seattle: University of Washington Press, 1995). All present excellent discussions of evolving forest use and policies in the Pacific Northwest.

3. Andrew Fisher, "'This I Know from the Old People': Yakama Indian Treaty Rights as Oral Tradition," *Montana* 49, no. 1 (1999): 2–17, provides an excellent example of how different those understandings could be.

4. Ruth Kirk, *Sunrise to Paradise: The Story of Mount Rainier National Park* (Seattle: University of Washington Press, 1999), outlines current geological, ecological, and historical understandings of the park through text and photos. (Ranger Aubrey Haines appears on page 103 releasing a campground "nuisance bear" from a livetrap in the 1950s.)

PREFACE TO THE
ORIGINAL EDITION

For many years I have been an admirer of Mount Rainier, the great rock-ribbed, snow-capped peak which is ever upon the horizon of those who dwell near Puget Sound. That attachment sharpened my interest in the mountain's early history and led me to a realization that the story of those interesting, adventurous times had never been adequately told, despite their great significance to the people of Washington State and to the nation as a whole.

And so I decided to attempt a re-creation of the events of that formative period which ended with the establishment of Mount Rainier National Park. Viewed with the perspective of the intervening years, those events form a sequence from discovery, through exploration and conquest, to the development of a public interest which resulted in a recognition of the mountain's national park character. Throughout, the unifying thread is something which has been aptly called "the mountain fever" — a contagious enthusiasm capable of driving some men into danger and hardship, capable of unleashing violent emotions in others, and capable of goading a few into altruistic accomplishment.

Readers will soon discover that the early history of Mount Rainier is so closely associated with the growth of organized mountaineering in the Pacific Northwest, that it is not possible to write about the one and ignore the other. However, I wish it to be understood from the beginning that this book is not a history of mountaineering, and there has been no attempt at thoroughness in that regard.

It will also be apparent that this is a serious work, intended for something more than mere entertainment. Thus, the trivia are deliberately included with the more important facts, in order to retain something of the spirit of the times in a work which must be devoid of literary license. Yet there is no reason why the reader should not use *his* imaginative powers; so I, like Shakespeare's supplicating chorus, suggest,

> . . . 'tis your thoughts that now must deck our kings,
> Carry them here and there; jumping o'er times,
> Turning the accomplishment of many years into an
> hour glass . . .
>
> *Henry IV.*

Aubrey L. Haines

Yellowstone Park, Wyoming
September 27, 1961

ACKNOWLEDGEMENTS

The writing of a history of bygone days is a cooperative venture in which the author is but the formalizer and interpreter of events made known to him through published and unpublished materials. He is thus indebted both to those who contemporaneously recorded the events and to those who make the records available.

The first group—those writers, compilers and editors of yesterday—are ordinarily acknowledged in appropriate foot-notes, and I have followed that practice. However, there is one whose work deserves more than perfunctory recognition. Fay Fuller's enthusiasm for mountaineering led her to col-lect and publish all she could find concerning mountain climbing on Mount Rainier. Without the "Mount Tahoma" stories and the miscellany of "Mountain Notes" published in her columns in the Tacoma *Every Sunday*, and later in the *Tacomian*, much of the early history of the mountain would have been forever lost. She did another signal service in keeping a scrapbook which has proven invaluable in filling in chinks in the record of mountaineering. Fay Fuller was the first woman to ascend Mount Rainier, but that feat, for which she became famous, is inconsequential in com-parison to her literary accomplishments.

The work of former Park Naturalist Robert N. McIntyre in compiling his *Short History of Mount Rainier National Park* has been of material assistance to me in the preparation of the chapters dealing with the legislative efforts to estab-lish a park, and I have followed him closely in that regard.

Among those to whom I am indebted for direct assistance in the course of my researches are two former librarians of the Washington State Historical Library at Tacoma, Miss Alta F. West and Miss Ruth M. Babcock; both were helpful far beyond the requirements of their positions. Others whose assistance was more than routine were Miss Priscilla Knuth, research associate and associate editor of the Oregon Historical Society; Mr. Gene Bismuti, assistant reference librarian of the Washington State Library; Mr. Donald G. Onthank of the Library Committee of Mazamas; Miss Mildred Dean, librarian of the Snohomish Public Library; Mr. Willard Ireland, librarian and archivist of the British Columbia Provincial Library; and Mr. Francis P. Farquhar, editor of the *American Alpine Journal.*

In addition to the libraries already mentioned, the collections of others were of material assistance, particularly the Mount Rainier National Park Library, the University of Washington Library, the Tacoma Public Library, and the Portland Public Library.

Many of the illustrations appearing in this work were prepared from photographs donated to Mount Rainier National Park; the assistance of Chief Park Naturalists Merlin K. Potts and Ross Bender in providing copies of some, and in allowing others to be copied, was most helpful. The kindness of Mrs. Maude Shaffer, who loaned valuable Longmire family photographs for duplication, is also acknowledged.

I wish to express again my sincere appreciation to the Western Imprints Fund of the Oregon Historical Society, to the Publications Committee and to Thomas Vaughan, director and general editor of the Society, for the publication of this study.

As always, my wife Wilma has been a willing and invaluable assistant to my labors, despite the demands of a growing family.

A. L. H.

PROLOGUE

This drama properly begins in the year 1833, but the stage was set forty-one years earlier when the misty coast of Drake's New Albion was at last outlined upon the map of the world.

And so we go back to Port Discovery in the year 1792, to the anchorage of His Britannic Majesty's sloop of war *Discovery*, and its consort, the armed tender *Chatham*, lying fog-shrouded and expectant near the waters of an inland sea yet unexplored. At five o'clock that Monday morning, the seventh of May, three armed and provisioned smallboats under the personal command of Captain George Vancouver were cast off to slip quietly away "for the purpose of becoming more intimately acquainted with the region in which we had so very unexpectedly arrived."

The little flotilla rowed eastward through the mists, centered in a ghostly circle of smooth, gray-green water where the softly liquid sounds of passage were muffled by the cadenced thump and creak of dripping oars. As they neared a point of land about noon, the fog dispersed allowing them a grand view of a broad inlet beyond which were snowy mountains with Mount Baker, already named for a lieutenant of the *Discovery*, at the north, and "a very remarkable high, round mountain, covered with snow, apparently at the southern extremity." Thus, did white men see for the first time the great volcanic peak which, on the morrow, would receive the name of Mount Rainier.*

---

*For May 8, 1792, Captain Vancouver wrote: "The weather was serene and pleasant and the country continued to exhibit between us and the eastern snowy range the same luxurious appearance. At its northern extremity Mount Baker bore by compass N. 22 E., the round snowy mountain now forming its southern extremity and which, after my friend Rear-Admiral Rainier, I distinguished by the name of Mount Rainier, Bore N. [S] 42 E." *A Voyage of Discovery to the North Pacific and 'Round the World* (London, 1801), II: 79.

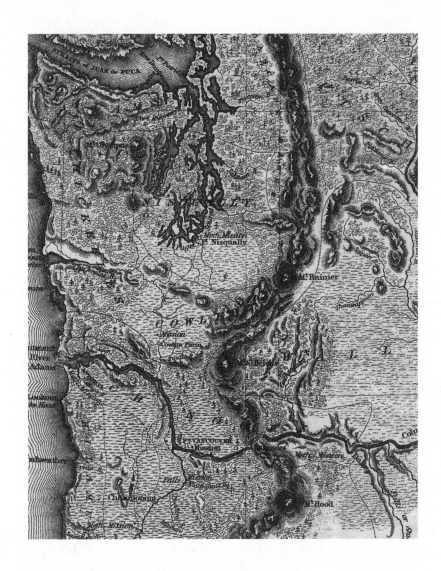

Wilkes map, 1841

CHAPTER I

WANDERERS IN THE WILDERNESS

1833-1857

*In 1833 the Puget Sound country was a wilderness
through which the first overland expedition of Hudson's
Bay Company men, under James McMillan, had passed
less than nine years before. It was a wilderness without a
white habitation until the previous year, when Archibald
McDonald selected the site for a trading post on the
gravelly plain of the Nisqually River, lodging there, in a
vast and lonely land, a tenuous foothold of civilization—
three white men and a little log building known as
"Nisqually House."*

THE HUDSON'S BAY COMPANY supply ship *Ganymede*
anchored in the Columbia River early in May, after a weary
voyage of more than seven months from London. Among
its passengers was Dr. William Fraser Tolmie, a young
Scotsman not long out of Glasgow University. He was glad
to be ashore at Fort Vancouver, but he was allowed little
opportunity to enjoy its rugged comforts, for he was immediately assigned to a northern station, Fort McLoughlin,
then building on Millbanke Sound in later British Columbia.

The little schooner *Vancouver*, which was to take him to
his post, had cargo for Nisqually House; so the doctor was
allowed to travel that far with an overland party under
Archibald McDonald. It was a welcome relief from the
cramped tedium of sailing ship life.

McDonald's party reached Nisqually nearly two weeks
ahead of the *Vancouver*, which made a slow passage from
the Columbia to Puget Sound. Before the vessel arrived, a

3

French-Canadian by the name of Pierre Charles split his foot with an axe while working on the new store building. The prompt and skillful attendance of Dr. Tolmie saved the man's life, but his condition remained critical on June 14, when the following entry was made in the Journal of Occurrences: "Ship getting in balast & water — Today it was [necess]ary to come to a decision respecting the professional attendance of Dr. Tolmie . . . His baggage is therefore landed and he remains here for the summer."[1]

The doctor was stranded at a rude, unfinished fort on the hot Nisqually plain, in sight of a mountain he did not fail to admire.[2] Though his diary speaks for his appreciation of the great peak which seemed, on clear days, to float detached above the blue outline of its foothills, it gives no indication whether his decision to visit Mount Rainier was the result of careful deliberation by the bedside of his patient or a matter of sudden resolution. It is only known that on August 27 he "obtained Mr. Heron's consent to make a botanizing excursion to Mt. Rainier for which he has allowed 10 days."[3] The purpose of the trip was given as the collection of "herbs of which to make medicine."[4]

Taking prompt advantage of his dispensation, Tolmie left Nisqually House two days later, with five Indians and three horses. His companions were Lachalet, the hereditary chief of the Nisqually tribe, and his nephew, Lashima; Nuckalkut, a Puyallup Indian, thought by Tolmie to be "a native of Mt. Rainier," with his relative, Quilniash, and an Indian whose name is unknown. Tolmie admits that the fees he offered in blankets and ammunition were less of an inducement to the Indians than the chance of killing elk and goats; indeed, before leaving for the mountain, Lachalet proceeded, with childlike enthusiasm and certainty, to sell and give away the fat he expected to get.

Mid-afternoon found the little party on the way, Tolmie riding an iron grey stallion and the Indians "disposing of themselves on the other two horses — except one who walked." In that manner they covered eight miles before sunset, camping for the night under a large "pine" at the abode of Nuckalkut's father. The old Indian had made his home in a beautiful, grassy opening among the oak trees,

4

and from it there was a view of a little water lily covered lake nearby, and far glimpses southward across the plain to the Nisqually River.[5]

A supper of sweetish, purple salal berries was followed by a night made uncomfortable by a drizzling rain. Toward morning, Tolmie was aroused from fitful sleep by the fall of a large, decayed branch, which gave him such a painful bruise on the thigh he "thought a stop was put to the journey." But he felt better after exercising.

By sunrise they were moving again, stopping at a marsh to breakfast on bread, salal, dried clams and a bit of dried goat meat after a warming, though damp, ride. The course was more easterly, bringing them, shortly after noon, to some rude shelters occupied by three Klickitat families, who met them "rank and file at the door to shake hands." Those flimsy bark sheds were well supplied with dried elk meat, some of which was immediately taken down and put in two kettles for the refreshment of the travelers. After a friendly smoke with their hosts, they made a "savage repast," then traded four musket balls and three finger rings for some dried meat to take with them.

The journey was continued in a heavy shower, along a route which became rougher and more beset with dense and tangled thickets as they approached the Puyallup River at a point between the present towns of McMillan and Orting. There they crossed to the east side on the horses, with Lashima carrying the baggage on his head; then they turned up the steam "through a rich alluvial plain . . . covered with fern about 8 feet high in some parts."

After passing through woods and crossing the river several times, the horses and equipage were left beside the stream at seven in the evening,[6] and the trip was continued afoot toward a house Nuckalkut knew of. But the house was found deserted, and probably in ruins, for they preferred to encamp on a dry bar beside the rapidly flowing river, described by Tolmie as "about 10 or 12 yards broad. Its banks are high & covered with lofty cedars & pines — the water is of a dirty white colour, being impregnated by white clay."[7] And there, as they sat by a cheerful fire, Lachalet tried to dissuade Tolmie from going on to Mount Rainier.

5

Tolmie slept well that night and arose to breakfast on two salmon caught by the Indians. Before leaving the campsite, Quilniash stuck the gills and the "sound" of the fish on a stick by the fire as a sign that salmon could be caught there.

That day the route paralleled the river through a dense forest,[8] from which they escaped at last by ascending the stream bed for several miles. It was rough traveling for Tolmie, who did not stand up well under the "smart trot" of the Indians. He described himself as "very inferior to my companions in the power of enduring fatigue." A night encampment was made on the river bank beneath some trees which gave them partial shelter from the incessant rain; from it they had a view upstream toward two lofty, wooded hills.[9]

Six o'clock on the morning of September 1, 1833, found Tolmie huddled under the dripping trees, making a few entries in his diary before arousing his sleeping companions. The prospect was a discouraging one—they had food enough for that day only, and the river was too high to be fordable in either direction. The advisability of continuing toward the mountain was carefully considered while some dried meat boiled in a cedar bark kettle; suprisingly, the Indians were for going on.

Before starting, Tolmie dressed in the Indian fashion — in a green blanket without trousers. But that proved less handy than he had expected; so it was traded for Lachalet's capot, "which had been on almost every Indian at Nisqually."[10] It was another day of alternate plodding through woods sodden with the never-ending rain, and splashing through the shallows and overflows along the river, which ascended rapidly between high, flanking hills. That evening they found a cave-like shelter where the river had undermined a gravelly bank, leaving a roof of large stones held in place by tree roots.[11] There they had a poor supper of berries heated with stones in a kettle, and there Tolmie wrote: "Propose tomorrow to ascend one of the snowy peaks above."[12]

However, the ascent was not made directly from their uncomfortable encampment, for they kept to the river for another three miles "to where it was shut in by [an] amphitheatre of mountains & could be seen bounding over a lofty precipice above."[13] Such a description fits only at the forks of the Mowich River, and surely it was from there Tolmie began the ascent of which he wrote:

> Our track lay at first through a dense wood of pine but we afterwards emerged into an exuberantly verdant gulley closed on each side by lofty precipices. Followed gulley to near summit & found excellent berries in abundance. It contained very few alpine plants. Afterwards came to a grassy mound where the sight of several decayed trees enduced us to camp.[14]

After tea at that high camp near the head of Lee Creek, Tolmie "set out with Lachalet & Nuckalkut for the summit which was ancle deep with snow for ¼ mile downwards." When they arrived at the top of Mount Pleasant, it was shrouded in a mist which was occasionally dissipated by a fitful southwest wind. The brief periods of clearing were sufficient to acquaint Tolmie with his surroundings: he found himself on a ridge-like summit bearing northeast and terminating in abrupt cliffs in that direction, and he noted that the snow-dappled ridges on both sides of the Mowich terminated "in Mt. Rainier a short distance to E." Returning toward camp, a vasculum of alpine plants was "collected at the snow." Later, they were carefully packed by the light of the campfire, after which Tolmie noted in his diary, "shall turn in. Lachalet by his own request is to be my bedfellow."[15]

The doctor lay shivering and wakeful through a rainy night, twice finding it necessary to rouse his swarthy companion to rekindle the fire; but about dawn a shift in the wind brought a clear-up, with frost. Sunrise found Tolmie and Quilniash climbing back to the summit over firm, granular snow which sparkled brightly in the sun.

From the top, "Mt. Rainier appeared surpassingly splendid and magnificent," but unfortunately, the notes covering the observations Tolmie made that morning were not recorded until the evening camp was reached. There, upon a "woody islet on Poyallip," he wrote the following:

> Tuesday Septr. 3— . . it [Rainier] bore, from the peak on which I stood S.S.E. & was separated from it only by a narrow glen, whose sides however were formed by inaccessible preci-

7

pices — Got all my bearings more correctly today the atmosphere being clear & every object distinctly perceived. The river
flows at first in a northerly direction from the mountain — The
snow on the summit of the mountain adjoining Rainier on the
western side of Poyallip is contin[u]ous with that of latter, &
thus the S. Western aspect of Rainier seemed the most accessible,
by ascending through a gulley in its northern side you reach the
eternal snow of Rainier & for a long distance afterwards the
ascent is very gradual, but then it becomes abrupt from the
sugarloaf form assumed by the Mt. — its eastern side is steep
[and] on its northern aspect a few d[ead][16] glaciers were seen
on the conical portion, below that the mountain is composed of
bare rock, apparently volcanic [a glacier] which about 50 yards
in breadth reaches from the snow to the valley beneath & is
bounded on each side by bold bluff crags scantily covered with
stunted pines — Its surface is generally smooth but here and
there raised into small points or knobs or arrowed with short &
narrow longitudinal lines in which snow lay.[17] From the snow
on western border the Poyallipa arose — in its course down this
rocky slope [the glacier] was fenced in to eastward by a regular
elevation of the rock in the form of a wall or dyke which (at
the distance I viewed it at) seemed about 4 feet high & 400 yards
in length Two large pyramids of rock arose from the gentle
acclivity at S. W. [NW] extremity of mountain[18] & around each
the drifting snow had accumulated in large quantity forming
a basin apparently of great depth Here I also perceived, peeping
from their snowy covering two lines of dyke similar to that
already mentioned.

The successful conclusion of that remarkable expedition
was recorded in the Nisqually House Journal of Occurrences
with the brief entry, "Dr. Tolmie returned safe after collecting a variety of plants."[19]

Soon after his return the young doctor was sent north to
his assigned station. Though he returned to Fort Nisqually
in 1836, spending most of the following twenty-three years
there, he made no further attempt to explore the great
mountain at the base of which he had once stood, considering in the frosty dawn if it might be climbed.

The next man to fall under the spell of Mount Rainier
did not climb it either, though he certainly wanted to. In
1842, a squadron of United States naval vessels, commanded
by Lieutenant Charles Wilkes, entered Puget Sound in the
course of an extended cruise of exploration. A base for
surveying and reconnaissance parties was established west

8

of Fort Nisqually upon the hill overlooking the little creek called "Sequalitchew." From there, the work of improving and expanding the charts and maps of the Puget Sound country was vigorously pushed.

While traveling over the neighboring prairies, Wilkes was so impressed by "the splendid appearance of Mount Rainier," that he determined to attempt an ascent if his duties would permit it.[20] There was no opportunity to gratify that desire, but his interest led to an attempt to determine the elevation of the summit by angulation. He says: "The height of Mount Rainier was obtained by measuring a base line on the prairies, in which operation I was assisted by Lieutenant Case, and the triangulation gave for its height, twelve thousand three hundred and thirty feet."[21]

It is probable that failure to correct the errors resulting from curvature of the earth, refraction of the atmosphere and height of the instrument station above sea level contributed greatly to the total error of the result,[22] though it is hard to believe that a naval officer who could be trusted with a squadron of ships was lacking in the fundamentals of geodesy, or the ability to apply them. Wrong though it was, the figure of 12,330 feet remained the official elevation of Mount Rainier until the United States Coast Survey raised it to 14,444 feet on the basis of careful angulation from the west. A minor consequence of Wilkes' error will appear later when considering a famous controversy over who was first to reach the summit.

While the Hudson's Bay Company drowsed on Puget Sound, metamorphosing into an agricultural company primarily concerned with farming and cattle raising on the gravelly plains at Nisqually, men began to appear who were of a different type from Doctor Tolmie or Lieutenant Wilkes. The first of those restless Americans who would soon crowd the Company from its feudal domain, came to Puget Sound in 1845, to be followed by other immigrants who brought the population of northern Oregon (that part of the Oregon Country north of the Columbia River) to slightly more than one thousand white people in 1851.

But there was dissatisfaction with the established government. At the Fourth of July celebration held that year in

Olympia (which developed from the earlier settlement of New Market), a suggestion was made to separate from the parent territory. Among the many grievances of the settlers was a feeling that the Oregon legislature had failed to provide them with the roads so essential to the settlement and economic development of the area. Roads *were* needed, particularly a wagon road across the Cascades to bring immigrants directly to the Puget Sound country from the Hudson's Bay Fort Walla Walla.

A road across the mountains (avoiding the tedious detour down the Columbia River, up the Cowlitz, and then overland to Olympia) became an obsession with the settlers on Puget Sound; so it is not surprising to find the second issue of the pioneer newspaper, the *Columbian,* printing an item of great interest to those readers of more than one hundred years ago. Under the simple title, "Visit to Mt. Ranier [*sic*]" the editor describes what may have been the first ascent of the mountain by white men.[23] According to Bancroft,[24] they were scouting the Nisqually Valley for a trans-Cascade wagon route, which leads to a presumption that they climbed the mountain's southern flank; but here is how the editor reported their trip:

About four weeks ago, a party of young men, consisting of Messrs. R. S. Bailey, S. S. Ford, Jr., and John Edgar [and B. F. Shaw], undertook an expedition to Mt. Ranier, for the purpose of ascending that mountain as far as circumstances might warrant. Ranier, as all are aware, is situated in the main Cascade range, distant from its base to Olympia about fifty-five miles. On arriving at the foot of the mountain the party secured their animals, and pursued their way upward by the back-bone ridge to the main body of the mountain, and to the heighth, as near as they could judge, of nine or ten miles[25]—the last half mile over snow of the depth of fifty feet, but perfectly crusted and solid. The party were two days in reaching their highest altitude, and they describe the mountain as extremely rugged, and difficult of ascent; on the slopes and table land they found a luxuriant growth of grass, far exceeding in freshness and vigor any afforded by the prairies below. On some of these table lands they found beautiful lakes[26]— from a half to a mile in circumference — formed from mountain streams, and the melting of snow. The party remained at their last camp, upward, two days and nights, where they fared sumptuously on the game afforded by the mountain, which they found very numerous, in

10

the shape of brown bear, mountain goat, deer, etc., with an endless variety of the feathered genus; the side of the mountain was literally covered with every description of berries, of most delicious flavor.

The party had a perfect view of the Sound and surrounding country — recognizing the numerous prairies with which they were familiar, to which were added in their observations, several stranger prairies, of which they had no knowledge, and which, probably, have never been explored. The evenings and mornings were extremely cold, with a wind strong and piercing — the noon-day sun oppressively warm.

They describe their view of the surrounding country and scenery as most enchanting, and consider themselves richly rewarded for their toil in procuring it. This is the first party of whites, we believe, that has ever attempted to ascend Ranier.

Not being provided with instruments for taking minute observations, and there being a constant fog and mist along the range of mountains, the party were unable to make any satisfactory discoveries in relation to a practicable route across them; yet Mr. Ford informs us, that he noticed several passes at intervals through the mountains, which, as far as he could see, gave satisfactory evidence that a good route could be surveyed, and a road cut through with all ease.

Who can calculate the benefit . . . to the Puget Sound country, had its citizens taken sufficient interest in the project to have located a road . . . let the people on the Sound be true to their interests the coming year, and turn their attention as early next spring as practicable, in surveying a route and establishing a road across the Cascades . . .

Who were those first white men to climb on Mount Rainier? It is probable that the leading spirit was John Edgar, a former Hudson's Bay Company shepherd married to a Klickitat woman. With in-laws across the mountains, he would have heard about the old Indian trails, but more important, he was an energetic, public-minded man. The scanty records indicate he was one of the twenty-six delegates to the convention at the Jackson house in 1851, in attendance at the Cowlitz Convention in 1852, and selected as a road viewer to locate the Naches Pass road in 1853. His public services were brought to an untimely end by his death in the Indian war of 1855-56.

Sidney S. Ford, Jr., was a son of one of the first American families to settle north of the Columbia River. At the time

11

of the ascent of Mount Rainier he was twenty-three years of age; stalwart, broad-shouldered, and kindly. He was already well-known for his experience following the wreck of the *Georgiana* on the east coast of Queen Charlotte's Island the previous year, and he would live on through enough adventures to make him a legendary figure in his own time. Settling later on Ford's Prairie, near Centralia, he reared a large family and died there in the year 1900.

Robert S. Bailey has enjoyed the honor of being thoroughly confused with George B. Bayley, a later climber. The extent of the confusion is indicated by a rather facetious newspaper item suggesting a "Move to Recognize Virtually Anonymous One."[27] While not as well known as his companions Bailey is *not* an unknown. He settled on Whidby Island in 1851, moving to Olympia the following year, where he became the assessor of Thurston County soon after its organization.

The name of Benjamin Franklin Shaw has been added to those listed in the *Columbian*, on the authority of Leslie M. Scott and the reminiscences recorded by George Himes.[28] The latter says:

> In this connection it is proper to state that in a recent interview with Colonel Shaw . . . I learned that he, with Sidney S. Ford and a man named Bailey, made the ascent of Mount Rainier in August 1854. While Colonel Shaw does not claim to have been on the highest point, he does say that 'no other point seemed higher than the one whereon his party stood.'

Taking into consideration that Shaw was seventy-seven years old at the time of the 1906 Oregon Pioneer Association meeting, where he talked with Himes,[29] and that he was relating an event which occurred fifty-four years earlier, it is not surprising he missed the date by two years. The fact that he lists two of the men known to have taken part in the 1852 ascent, and agrees on the month, is strongly suggestive that he was one of the climbers.

The possibility that Shaw's reminiscence is merely an attempt to claim credit for an accomplishment he knew of, but did not participate in, is not admissible. "Frank" Shaw was one of the most respected and capable men of the formative period of Washington Territory. He was a mem-

12

ber of the first party of Americans to push into northern Oregon in 1845; a tall, strong lad of sixteen, with flaming red hair and some knowledge of the millwright's trade. In partnership with Simmons and Bush, he built the first grist-mill on Puget Sound in 1846, then served with the Oregon Volunteers during the Cayuse War of 1847. After the war, he engaged in the lumber business until the arrival of Isaac I. Stevens, first governor of Washington Territory. Stevens needed an interpreter to assist him in making treaties with the Indians, and Shaw was recommended because of his outstanding ability in the Chinook jargon. Universal respect for his fairness and good judgment led to an appointment as a lieutenant-colonel of Washington Volunteers during the Indian war of 1855-56, and to him belongs much of the credit for the successful conclusion of that conflict. Such a hero, with glory to spare, would hardly insinuate himself into a relatively minor event.

Returning to the Cascade road, the *Columbian* continued its vigorous demand for action.[30] In April of 1853, the public was asked to subscribe money so that a route could be opened, and in May came the chiding demand, "down with your dollars!" Then, in June, they were wisely admonished, "prepare for the immigrants . . . sell not one dollar's worth of provisions to be exported from the country — husband well your resources . . . that you may be prepared to extend true, substantial, and thrice hearty welcome to the wayworn new comers."

During the following months, money and services were subscribed to the value of $6,000. John Edgar traced out a route along the old Naches Pass trail used by the Yakima and Klickitat Indians in crossing between their country and Fort Nisqually, and work was begun in August at both ends of the route. The first "tourist" passed that way as the crews worked. He was Theodore Winthrop, author of *The Canoe and the Saddle*. By applying the name "Tacoma" to Mount Rainier, he unknowingly furnished the name for a city, the substance of a mighty dispute, and assured his own immortality[31]— quite an accomplishment for one who was making a hasty ride over a less than half finished right-of-way!

Edward J. Allen, who has been called the "engineer,

13

contractor, and soul of the work,"[32] pushed the road with such vigor that enough was constructed to permit the first train of thirty-five wagons to cross the mountains that October. In it was a man by the name of James Longmire, from Fountain County, Indiana, who has left a vivid description of the arrival on the Nisqually plain:

> On the 10th of October, Mr. Tolmie, Chief Factor of the Hudson's Bay Company, stationed at Fort Nisqually, paid a visit asking us numerous questions about our long journey and arrival, treated us in a very friendly manner but soon left after bidding us a polite farewell. In about three hours he returned with a man driving an ox cart which was loaded with beef, just killed and dressed, which he presented to us, saying, "Distribute this to suit yourselves." We, not quite able to understand the situation, offered to pay for the beef but he firmly and politely refused, saying, "It is a present to you."[33]

And with the beef they were given notice not to settle north of the Nisqually; so the families were left encamped on the plain while the menfolk crossed the river in search of homesteads on the prairies to the south. James Longmire found himself a place on the Yelm Prairie, beyond the lands grazed by the sheep and cattle of the Puget's Sound Agricultural Company. But there were other Americans with less respect for the Company's interests; among them, William Packwood, who had settled on the fertile Nisqually bottom in 1847, to become a fractious neighbor and a thorn to the kindly doctor.[34]

During the month of March in that same year of 1853, a young second lieutenant by the name of August V. Kautz arrived at Fort Steilacoom with Company C of the Fourth Infantry Regiment. He liked that isolated post where "there was plenty to eat, and little to do, and pleasant surroundings."[35]

The fort occupied the site of a former sheep ranch, leased from the Puget's Sound Agricultural Company by an English settler named Joseph Heath. After his death, twenty acres of land with the buildings, were rented to the United States for the use of troops sent to garrison the Puget Sound country. As Kautz knew it, the fort consisted of "a collection of buildings built in the shape of a square, the men's barracks on one side, the officers' on the other, storehouses on another and a row of army wagons on the last side."

From his quarters beneath the great oak trees, Lieutenant Kautz had a splendid view across the parade ground toward Mount Rainier. He was first fascinated, then enthralled by the great snow-covered, unknown mountain and he soon determined to conquer it. But Kautz was sent to southern Oregon that winter to take part in the Rogue River Expedition,[36] leaving his mountain to others for a time.

Here the story depends entirely upon the reminiscence of an aged Indian who claims to have guided two unknown white men to Mount Rainier, where they ascended to the summit from the north side. Nothing has come to light to confirm any part of the tale, yet it invites credence for there is a ring of authenticity to both versions of the ascent of 1855.[37]

The Indian who is the narrator of this account is listed in the official files of the Yakima Indian Reservation as Saluskin (no first name, though easily identified by adding "the chief"). He was born in 1823 and died December 30, 1917.[38] Alec Saluskin, his son and a member of the tribal council, says his father was a tall, strong man, keenly interested in hunting and horse racing. He was a consistent friend of the white settlers in the Yakima Valley and a successful stockman, with many cattle and horses, but, like many of the older Indians, he never learned to speak English. Here in his story, as related to Lucullus McWhorter in November, 1916:

It was, I think, one or two years after this,[39] our people were camping above the Moxee Bridge.[40] For a long time a big topis [pine tree] stood there.

One day an old man, Ya-num-kin, came to me and said: "Two King George men come." I look out and see them; both short [not yet] middle age.

They come to us; one a short man. Black eyes like Indian. Fine looking man. Clean face. Some old Indians said: "He is Mexican." His clothes look like corduroy. He wore a hat and had a big, banded flint-lock pistol. It shot big bullet.

The other man, tall, slender, not good looking, but about right. He had brown, not quite red hair on upper lip; light hair and brown eyes. He looks some mixed blood with white,

15

just a little mixed. He had grey clothes and cap. He had long flint-lock musket. Shot big ball and buckshot. I found the short man had the strongest mind.

They rode Indian horses, one blue, or roan. Had two pack horses, one a buckskin. No big or American horses. Here all cayuses. No white men here; old man Thorpe had not come.[41]

They wanted to know a man who could go to the White Mountain. The old people were afraid and said: "Do not show them the trail. They want to find money" [mineral]. The Indians asked: "Why do you go to the White Mountain?" The man said: "We are Governor Stevens boys. We come up the river from Walla Walla and look for reservation line made at treaty." They had long glass to look through.

Then the old people said: "All right." They told me to show them the trail. I am old man Sluskin now. I was young then. My father[42] raised me here; I knew the trail. I asked my father if I must go. He said: "Yes!" I was not afraid. It was about the middle of June and patches of snow still in mountains.

I started, leading the buckskin packhorse and my extra saddle horse. I took them to mouth of Tieton and camped. We got lots of trout; plenty of fish.

Next day we travelled and camp at night in Tieton Basin.[43] The men catch plenty of fish again.

Next day we went to Ai-yi [Fish Lake] and camped. We camped at mouth [head of Bumping River] at head of lake.

We go on to big ridge near head of Natches and camped. The next morning the men looked with glass every way.

Then we started and went to Tahoma the big White Mountain. They look all around. South side is bad. They ask me about west side. Yes, I know it. On sunny side [east] water comes out; called *Mook-mook*. Dirty water from the middle of mountain and ice.[44] The tall man killed a young *yamas* [fawn] as we crossed the *Mook-mook*. Shot it as it ran in front of us. This was all the game killed.[45]

We got to ridge like place and found plenty green grass and nice lake, good size called *Wa-tum*. We camped there.[46] The men looked every where with glass.

The *Sum-sum* [sharp ridge] runs down from the mountain. It was covered with *wou* [goats].

The men asked me if I could catch sheep for them. I tell them no; only when have young ones. They said: "If you catch one, we will buy it, big one." I never try to catch that sheep; too wild. That night we roasted *yamis*.

Next morning we go to a lake, not a big lake — only *tenas* [little] big, at foot of mountain. We got there about one hour after noon, camped and had dinner. This was north side of mountain.[47]

16

Next morning the men took glass up side of mountain and look. They asked if I could take them to top of mountain. I do not know the trail. Too many splits in ice. We camped overnight and roasted *yamis*. The men said: "In morning we go somewhere."

Next morning I see them put lunch in pockets and leave camp. I did not know where they go; but they start up the mountain. They put on shoes to walk on ice. No, not snow shoes; but shoes with nails in two places like this [touching heel and toe]. They start early at day light and come back at dark, same day. I staid in camp all day and think they fall in ice and die. At night I see smoke go up at top of mountain and I hear it like low thunder. They did not tell me if they hear the thunder.[48]

The men told me they went on top of mountain and look with glass; along Cascades, towards Okanogan and British Columbia: Lake Chelan and every where. They said: "We find lines." They tell me they set stick or rock on top of mountain. I did not understand much Chinook and could not tell if ilquis [wood] or rock.

They said ice all over top, lake in center and smoke or steam coming out all around like sweat-house.

Next day I went home[49] and did not know where these men went. I left them there. I do not know if they got other Indian to guide. Before I started, each man gave me a double blanket and shirt. They gave me cotton handkerchief, big and green striped. A finger ring, lots of pins and fish hooks. *Too-nes* [steel], *sow-kus* [flint] to make fire; a file and hatchet. They gave me a lunch of *yamis*. I was two days and a half getting home.

On this trip, concluded the chief, I tasted bread for the first time. It was nice. We had no coffee; some kind of tea made from berries.

. . . if you do not understand my talk: if not interpreted straight; then you will write it as a lie. It must be right. Chinook is not good for story.[50] I am glad to have two interpreters.

White people are always making me stand up and talk. Why is this? I do not understand what they want. They get me tangled. Then the *temis* [newspaper] tells my talk different from my words.[51] I do not want this. It is a lie. It is same as stealing.

That is Saluskin's story of what may be the first complete ascent of Mount Rainier by white men. In referring to the narrator, his name is used as his descendants now spell it; however, there are many variants. The Yakima Agency records also list him as "Sluse-cum," and "Salooskin," A. J. Splawn refers to him as "Shu-lu-skin," and the newspapers universally called him "Sluiskin," which has led to no end of confusion between the chief and another Indian who later acted as a guide for Stevens and Van Trump at the time of their ascent of Mount Rainier. When asked if he was related to the Sluiskin who guided in 1870, Saluskin answered: "I am no relation to Shluskin with the crippled hand. He was half-sister on the father's side, to my wife . . . He drowned in the Yakima River several years ago. Never found his body. I never heard he took two men to the White Mountain."

Sluiskin, the guide of 1870, has been well honored on the map of Mount Rainier National Park. His name has been given to the waterfall beside which he awaited the return of Stevens and Van Trump from their successful ascent, and to a mountain on the north side of the park; while Saluskin, the guide of 1855, has been entirely neglected. It would be very appropriate to rename the Sluiskin Mountains in honor of the Indian guide who waited at Mystic Lake while the two unknown white men climbed to the summit of Mount Rainier more than one hundred years ago.

Following the signing of the treaty at the Walla Walla council, war clouds gathered ominously over the Yakima and Klickitat country, casting shadows of unrest upon the tribes west of the mountains. The causes of the Indian war of 1855-56 are of no concern here; it is sufficient that the storm broke with the murder of Agent A. J. Bolon by Yakima Indians late in September 1855.

An initial defeat suffered by the small force of regular troops under Major G. O. Haller left the settlers in the Puget Sound country to face a war for which they were ill-prepared. There were less than 1,600 men capable of bearing arms, to augment the three small garrisons of regular soldiers. Many of the settlers lacked guns suitable for military service; indeed, the volunteers hastily mustered into the service of Washington Territory could not have been adequately armed without the help of the Hudson's Bay Company.

Upon the scattered settlements, distraught with rumors, poorly prepared and lacking the guidance of their able governor (then on a treaty-making trip to the Blackfoot country), war finally came late in October. Indians from across the mountains were joined by local dissidents, under the Chiefs Leschi and Quiemuth, in a savage onslaught. It was a war of ambush and murder, of shots fired into lighted windows, of sudden flight from burning homesteads, with all the attendant suffering and anguish.

Many settlers survived because the warnings of friendly Indians gave them time to seek the protection of block-house or town. The Longmires were among those fortunate ones; an old Indian by the name of "Stub" came to the house and told them war was imminent, so that James had time to place his family in safety. After the outbreak, he joined Captain Charles Eaton's company of Puget Sound Rangers.[52] William Packwood became the sergeant of a guard of ten men at the vital ferry over the Nisqually River between Olympia and Steilacoom, while "Frank" Shaw enlisted a company of men to go to the rescue of Governor Stevens and conduct him and his treaty-making party (which included the governor's thirteen-year-old son, Hazard) back to Olympia.

Of course, Sidney Ford, Jr., fitted himself into the scene with his customary boldness. After brief service with the Stevens Guards, he undertook the leadership of a force of Indian allies used to scout beyond the fringes of the settlements. On one such expedition he lay at night pretending sleep while his "friendly" Chehalis Indians held a long discussion on the advisability of killing him and joining the hostiles.

Scouting *was* dangerous business. A force of fifty regulars and as many volunteers, with John Edgar and a man named Perrin as scouts, was sent against Indians known to be on South Prairie. Reaching South Prairie Creek a little in advance of the troops, the two scouts were shot by Indians while crossing on a log. Litters made of boards torn from a small cabin were used to carry the wounded men five miles through the forest to camp on Connell's Prairie "under cover of darkness, with nothing to light our path, (and

19

dared not use it if we had)."⁵³ John Edgar's wound was mortal, and the territory lost a useful, public spirited citizen — a man of good works.

Just when the prospect seemed darkest, the Oregon Volunteers courageously campaigned against the Indians east of the mountains. Though inconclusive, the effort had two beneficial effects: the way was opened for Governor Stevens to return safely, and the Indians who had come across the mountains to stir up war on Puget Sound returned to defend their country. Thus, left without support, the hostile Indians of western Washington broke into small bands which could do no more than raid the settlements from well-hidden retreats. There was now an opportunity to end the conflict.

In February of 1856, Lieutenant Kautz returned from southern Oregon to take part in a campaign against the hostiles in western Washington. He was wounded in the leg while leading a spirited charge against the Indians during the battle of White River on March 1, but was able to take the field again in April, to lead a scouting detachment of fifty men to the foothills west of Mount Rainier. His regulars brought back about thirty prisoners, men, women and children, survivors of an attack by Captain Maxon's volunteers upon their encampment on the Mashel River. The captives were well treated, and some were later sent out to induce the remaining hostiles to surrender, which they were glad to do. Thus, the war west of the mountains was brought to an abrupt conclusion.

There was yet much fighting to be done east of the Cascades, but we need be concerned with no more of it than the battle of Grande Ronde, where Lieutenant-Colonel Shaw, with a mixed force of Oregon and Washington volunteers, came upon hostile Indians in the open valley.

Although taken by surprise, they received him in a defiant attitude; large numbers of braves, mounted and armed, with a white scalp borne on a pole among them, confronted him . . . on Captain John's approaching them to parley, [they] cried out to shoot him . . . throwing off his hat, and with a shout, the tall, rawboned leader of the volunteers instantly charged at the head of his men, his long red hair and beard streaming in the wind, broke and scattered the Indians, chased them fifteen miles.[54]

One of the Indians who fled before the wrath of "Quatlich" was a young Yakima warrior by the name of Sluiskin. He was marked from that day on by his bullet-shattered hand.

It was some time before Chief Leschi, of the Nisqually tribe, surrendered at the urging of his old friend Dr. Tolmie. He was then charged with murder by civil authorities, because of his alleged part in the ambush slaying of two citizens, Joseph Miles and A. Benton Moses, early in the war. The public anger was so aroused against the chief that it was necessary to place him in the custody of the military at Fort Steilacoom. While awaiting trial and his subsequent execution, Leschi was guarded by Lieutenant Kautz, who came to believe Leschi innocent, and did what he could to assist the chief.[55]

The result was a close friendship between the young officer and his prisoner, who gave Kautz much valuable information concerning the route to Mount Rainier. He was even then considering an attempt to climb the mountain, and Leschi had some hopes of going along as guide; but the chief finally realized his execution was inevitable, recommending another Nisqually Indian, old Wa-pow-e-ty.

During the year and more following the execution of Leschi, Kautz's mountain climbing plans were not taken seriously. He says: "I had expressed so often my determination to make the ascent, without doing it, that my fellow officers finally became incredulous and gave to all improbable and doubtful events a date of occurrence when I should ascend Mount Rainier."[56] About the first of July, 1858, his resolution "took shape and form," and, on the sixth, there is this entry in his diary: "I was busy all day preparing to go up Mount Rainier. I find the men quite willing to volunteer to go."[57] So was Assistant Surgeon Robert Orr Craig, who arrived from Fort Bellingham the following afternoon.

Completing their preparations, which included making alpenstocks and ice creepers for each member of the party, in addition to gathering the necessary rations and equipment, they were able to make a start just after noon on July 8. Four soldiers of the 4th Infantry Regiment went with pack animals directly to Charles Wren's place,[58] beyond which there was only wilderness, while Kautz and the doctor rode by way of the Nisqually Reservation to get Wapowety, who agreed to be their guide. The night was spent at Wren's.

On the ninth, they followed the old Indian trail into the great forest that lay between the Nisqually Plains and Mount Rainier, passing by way of the little prairies which provided the only relief from the cathedral gloom of their route. First beyond Wren's was a camas prairie called *Pawtumni;* followed by another known as *Koaptil,* while the *Tanwut,* a small stream flowing from present Tanwax Lake, was crossed about two-thirds of the way from Wren's to Mishawl Prairie, which was as far as horses could be taken as well as the limit of Kautz's knowledge of the country. They camped there in a beautiful oval meadow in weather which had at last cleared.[59]

The next morning two soldiers, Bell and Doneheh, were left to care for the horses while the remainder of the party made an early start afoot toward Mount Rainier. Each one carried a blanket, a canteen (which the doctor foolishly filled with whisky), a haversack with two pounds of dried beef and two dozen crackers,[60] and his climbing equipment; while Pvt. Nicholas Dogue, a German soldier, carried the fifty-foot rope, and Pvt. William Carroll, an Irishman, brought the hatchet. Kautz also carried a field glass, prismatic compass, thermometer, spirit lamp, and large revolver. Wapowety had his gun, upon which the party depended for fresh meat.

Crossing the Mashel River, then at low stage, they climbed the ridge beyond in order to pass around the falls of the Nisqually. The top was reached at three o'clock, after wearying toil amidst underbrush and down timber, during which the doctor developed severe cramps in his legs and only managed to continue by hiring the little Indian guide to carry his pack the round trip for ten dollars.

As they rested at the summit, there was a cooling breeze and a grand view of the country to the west, with "nothing definite except forest — of which there was a great excess." It may have been there the doctor poured out his whisky in disgust and gladly took some water in its place.

They continued eastward along the ridge for four or five miles, searching the ravines for water, until their great weariness forced them to camp in the forest "without water except what we had in our canteens." It was already apparent their guide knew very little about the route.

On July 11, they at first veered more to the east in their course, in order to stay on the high ground, but that led them into a large basin; so they turned south to the Nisqually, through a "dark, gloomy forest, remarkable for large trees, and its terrible solitude," arriving at the river about three o'clock. There the traveling was no better, but they had water. In fact, they were wet much of the time from wading between gravel bars, and were glad to encamp upon the bank six miles up the stream. That day they had difficulty eating even the small ration of two ounces of beef and four crackers, though their strength depended on it.

The third day, which was Sunday, another early start was made with the route lying along the north side of the river through thick underbrush until late afternoon, when they passed through a burned area of about one hundred acres where blackberries were plentiful. The failing strength of the doctor, who was in poor condition and unable to eat, forced them to camp early at a point where the appearance of the opposite hills led them to believe there was a sizeable tributary entering.[61]

On the morning of the thirteenth, Wapowety killed a deer, the first animal seen since the "large red wolf" had scuttled quickly from sight two days before. All ate plentifully. The remaining meat was smoked, and they were on the move again by eleven o'clock. A narrowing valley brought them, by nightfall, to the west bank of the milky stream now known as Tahoma Creek where they encamped. From there, the mountain was in view, appearing very close, but the weather seemed to be changing.

When they awakened the following morning, the sky was overcast. They were in doubt whether to go up Tahoma Creek or the river, but finally decided on the latter. As they progressed, the canyon narrowed, turning toward the mountain, the stream became more rapid, and travel more difficult. Forcing their way through thickets, and repeatedly wading the cold torrents which poured down an inclined plane of gray boulders, they passed avalanche debris lying in a tumbled confusion of snow, trees, and rocks, beneath perpendicular cliffs extending into the clouds. Terribly wearied, they came at last to the "foot of an immense glacier, from which the River emanates," camping there under drooping tree limbs which gave them some protection from the drizzling rain that began with the night.

They were tired, dirty, hungry and dejected. While the others rested, Kautz hastily explored the foot of the glacier, making a crude sketch of it.[62] He also found the boiling point of their tea water to be 202° F. at that encampment.[63]

The morning of July 15 was foggy with rain which occasionally changed to snow or sleet. Despite that, the upward trek was continued. After some difficulty getting up the steep meltface of the glacier, they made their way along the slippery, inclined surface until noon, when the visibility decreased and crevasses were encountered. With considerable effort, and the use of their rope, they "crossed over to the moraine on the west side, finding with much difficulty, a camp among the pines."[64] The remainder of the day was spent resting in the shelter of the stunted trees. Toward evening Kautz climbed the ridge above their camp, deciding that it offered the best route for an ascent of the mountain (which he thought could be made in three hours).

While at that camp, whistling marmots were observed feeding around burrows where the excavated earth was covered by old goat tracks. A hasty conclusion that the tracks were made by the animals led Kautz to give the following description of them upon his return from the expedition:

> They saw numbers of mountain-sheep, a small animal with long shaggy black and whitish hair, with the appearance and attitude of a small dog, and the motion and feet of a sheep. They are exceedingly wild—burrow in the earth, and at the least alarm, make for their holes.[65]

That ludicrous mistake was the source of much embarrassment.

The boiling point of water, taken during the evening, was found to be 199° F., with the temperature standing at

34°. George Gibbs, who was attached to the garrison at Fort Steilacoom in the capacity of interpreter, later reduced the observation as follows:[66]

Altitude, Kautz's encampment, by Loomis's formula—
Barometric pressure, corresponding to 199° 22.971
Assumed sea-level 30.042
Assumed altitude 7011.4
Correction for temperature 234.0
                                                                    7245.4
Correction for decrease in gravity 23.0
                                                                    7268.4

However, the encampment could not have been above an elevation of 6,500 feet, which is the upper limit of the forest on that side of the valley.

Upon awakening the next morning, they found snow on their blankets, but the clouds began to lift at eight o'clock, allowing them to begin the climb to the summit. They were enveloped in clouds, with only occasional glimpses of the peak until noon, after which they passed into brilliant sunshine and noticeably cooler air. Below them the sea of clouds lay smoothly, "above which appeared the snowy peaks of St. Helens, Mt. Adams, and Mt. Hood, looking like pyramidal icebergs above an ocean." About two o'clock the clouds broke up, giving them a view of the surrounding country. It was a grand panorama which left a vivid impression of rocky peaks projecting through a dark mantle of forest relieved only by the winding, silvery ribbon of the Nisqually River.

Climbing was difficult. The snow was so porous in places that they sank to their knees. At four o'clock, the doctor, Carroll, and Wapowety dropped behind at about the 12,000 foot elevation, but Kautz continued, followed closely by Dogue. They finally reached, and passed, a difficult area of seracs and crevasses, beyond which the summit appeared to be. It was but an illusion, for they were only upon the edge of the prominent saddle between Point Success and the main summit. There Dogue threw himself down saying he could go no farther.

Kautz continued on alone for half an hour, reaching an elevation of 14,000 feet,[67] just as the sun was setting behind the Olympic Mountains on July 16, 1857; at that point he reluctantly turned his back upon the summit and returned to the place where he had left the soldier (without his hat which was carried away by a violent gust of wind when he was well upon the saddle). Kautz was surprised to find the doctor with Dogue. They held a hurried consultation, deciding it was impossible to remain on the mountain and live, for ice was forming in their canteens and they had no blankets; nor would it be possible for them to descend through the crevassed area if they delayed longer; so, reluctantly they began the descent shortly after six o'clock.

They found the downward journey easier, accomplishing in three hours what had required ten on the ascent. On one icy slope Dogue lost his footing and rolled thirty to forty yards, but he came through it with nothing worse than scratches. They reached camp well after dark to find Wapowety worried about them.

Kautz wished to make another try for the summit, in the hope of determining its elevation by an observation of the boiling point, but in the morning, the Indian's eyes were so inflamed he was blind. Also, a check of their provisions showed that they had but four crackers and a pound of dried meat each for the return journey to Mashel Prairie. Since Wapowety would no longer be able to augment that meager ration by his hunting, they dared not linger at the mountain.

The return was begun at once, with Kautz wearing a cap made from the sleeve of a red flannel shirt in lieu of the hat he had lost. While passing down the Nisqually Glacier, Kautz was able to make several interesting observations concerning it; in particular, he noted its size as "a mile wide and four or five long, probably half a mile deep,"[68] and he was impressed by the crashing and grinding noise it made both day and night. During the day they passed two of the camps made on the way to the mountain, stopping at dark where they had seen the blackberries in the old burn. A dead tree was set alight to assist them in gathering berries for their supper, but it afterward endangered them with its brands.

On July 18, they breakfasted on berries, then continued

26

down the river, slowed by the underbrush and their exhaustion. Passing the encampment of the evening of the eleventh, they traveled on to an old Indian campsite where they stopped, knowing they could improvise a shelter of the cedar bark scattered about, if the weather continued to worsen. The doctor fared poorly as he had been carrying his own pack on the return journey.

The rain did not occur and next day they continued down the river to a point five or six miles below where they first reached it. Wapowety was able to carry the doctor's pack, assisting him greatly. Kautz tells of listening to the soldiers discussing hunger at that camp where the last of the food was eaten. Carroll said: "I've often seen the squaws coming about the cook-house picking the pitaties out of the slop-barrel, an' I thought it was awful; but I giss I'd do it mesilf this mornin'."

They hoped to reach the camp on Mashel Prairie on the twentieth, but the march across the mountain proved harder than they expected. Their water was used up before noon and they arrived at the Mashel River too tired to do more than quench their thirst before lying down to sleep. Though but two miles from the camp where Bell and Doneheh waited for them, it was an impossible distance that day.

They were up and on the way early the next morning, but mistook the direction of camp, as they supposed they were below the point of their previous crossing of the Mashel, when actually two miles above it. Though some time was lost, they still reached camp before six o'clock to find Bell and Doneheh asleep.

Kautz ate sparingly (half a cracker, with butter and coffee), warning the others to do likewise, but they apparently failed to heed him. After a rest and a more substantial meal, he and the doctor saddled their horses and rode to Fort Steilacoom, where they arrived at dusk. The soldiers came in at their leisure.

Neither was readily recognizable; they were sunburned, haggard, and tattered, with Kautz wearing his shirt-sleeve cap, and the doctor a flour sack in place of a trouser leg. During the two weeks they were gone, both had lost weight. Kautz adds:

27

The two soldiers went into the hospital immediately on their return, and I learned that for the remainder of their service they were in the hospital nearly all the time. Four or five years after, Carroll applied to me for a certificate on which to file an application for a pension, stating that he had not been well since his trip to the mountain. The Indian had an attack of gastritis and barely escaped with his life after a protracted sickness . . . The doctor . . . was taken with violent pains in his stomach, and returned to his post quite sick. He did not recover his health again for three months.

Kautz also suffered for the rest of his life as a result of the rigors of that expedition. His diary entry for July 21, 1857, states: "I suffered all day from the Piles, for the first time in my life."

The prediction Kautz later made that "many a long year will pass before roads are sufficiently good to induce any one to do what we did in the summer of 1857," was really not a good guess, but his party *was* the last to force their way into that great wilderness without leaving a mark upon it. Their successors would blaze a route, cut a trail and build a road — there would be no more wanderers in the wilderness.

CHAPTER II

CONQUEST OF THE HEIGHTS

1858-1870

*After the bold and almost successful attempt of Lieu-
tenant Kautz and his party to ascend Mount Rainier there
appear to have been no further efforts toward its conquest
for the next thirteen years. They were years during which
the great Cascade wilderness surrounding the mountain
lay untouched, for the impoverished Territory of Washington was but partly recovered from its Indian war when
the war between the States and the lure of the gold mines
of Montana and Idaho drew its adventurous young men
away. Snowden says of Washington during that period
"the western part seemed hardly to progress at all."*[1]

In July of 1858, William Packwood and James Longmire
explored the old Indian trail up the Nisqually River to a
point which they called Bear Prairie from the animal killed
there. Afterwards they talked so favorably of the possibility
of building a wagon road by that route to the east side of
the mountains that the legislature passed an act locating a
territorial road across the Cascades south of Mount Rainier
from Yelm Prairie to the Naches River, appointing Packwood, Longmire, and G. C. Blankenship "to view out and
mark the same."[2] The following June, Packwood, Longmire
and W. Kirtley (who had explored on the east side of the
mountains with Blankenship the previous year) again went
up the Nisqually River intending to cross completely over
the Cascades by the route proposed for the new road. They
traveled on foot, carrying their food, blankets and equipment on their backs, for eight difficult days, reaching the
stream now known as Skate Creek, which they followed

eastward to its entry into the Cowlitz Valley, "fully satisfied that we had got to the opening of the Nachess Valley."

In that they were mistaken. The summit of the Cascades was fourteen miles to the east, and the valley they were seeking lay even farther — beyond a timbered jumble of ridges and drainages which could offer only a difficult route for a wagon road. Such a mistake arose naturally enough from a geographic misconception held by Puget Sounders of that day. They thought Mount Rainier stood astride the summit of the Cascades, not realizing it was positioned sentinel-like upon the west flank, with the valleys of the White and Cowlitz rivers curving in behind it like encircling arms.

And so, they turned homeward,[3] sincerely believing an easy road over the mountains had been found. It is fortunate for the reputation of those eager road viewers that the press of circumstances already mentioned prevented the territorial road they were promoting from ever being built. However, the blazes they left along the old Indian trail remained to mark the "old Packwood trail," and later the "Yelm trail," which would be *the* route to Mount Rainier for more than twenty years. But for another decade the trail was unused.

As the blazes along the old Indian trail lost their raw, fresh look and the down timber accumulated over its fading treadway, three men came to Washington Territory who would follow those blazes and use that trail in a spectacular conquest of Mount Rainier. General Hazard Stevens was the son of the first governor of Washington Territory.[4] He was born at Newport, Rhode Island, June 9, 1842, into the family of Lieutenant Isaac I. Stevens, who was the army engineer charged with strengthening the defenses of New Bedford. That able officer rose rapidly in the service by virtue of his ability and his gallantry in the Mexican War, but by 1853 he despaired of advancing beyond the rank of major, which he then held. And so, he asked for, and received, the governorship of the newly created Territory of Washington.

In 1854, Governor Stevens was able to bring his family out to the new home at Olympia, in the sparsely populated

30

territory through which young Hazard often accompanied his father on treaty-making trips among the Indian tribes. Both father and son volunteered for service with the Union army during the Civil War. At the battle of Chantilly, September 1, 1862, Major-General Stevens was killed and Captain Stevens was severely wounded in the same charge upon the Confederate troops of Jackson, Hill and Longstreet (a costly rear-guard action which saved from complete destruction a beaten Union army retreating from the second battle of Bull Run). Hazard Stevens recovered from his wounds to become the youngest brigadier general of volunteers in the Union army, and was mustered out of service with a record which brought him an award of the Congressional Medal of Honor — the nation's finest tribute to individual valor.

With the war well behind him, Hazard undertook to support his widowed mother by returning to Washington Territory in 1867 as a federal revenue collector. His job required him to ride horseback between the scattered settlements, giving him ample opportunity to admire the great peak which dominated the eastern skyline. Gradually he caught the mountain fever and determined to make an ascent, but he could not at once find a companion. When he did, he found the right one.

Philemon Beecher Van Trump was born at Lancaster, Ohio, December 18, 1839, and so was a little older than Stevens.[5] He came of New York Dutch stock and received an education at Kenyon College and the University of New York. In 1865 Van Trump crossed the plains to Montana and Idaho where he prospected for gold without success, continuing on to Portland, Oregon. From there, he traveled afoot to the Sacramento Valley, earned enough for a ticket East and returned home by way of the Isthmus of Panama in 1867. In the summer of that same year Van Trump came to Washington Territory as the private secretary of his brother-in-law, Marshall F. Moore, the seventh governor. He, too, caught the mountain fever. He says:

I obtained my first grand view of the mountain in August, 1867, from one of the prairies southeast of Olympia. That first true vision of the mountain, revealing so much of its glorious

31

beauty and grandeur, its mighty and sublime form filling up nearly all of the field of direct vision, swelling up from the plain and out of the green forest till its lofty triple summit towered immeasurably above the picturesque foothills, the westering sun flooding with golden light and softening tints its lofty summit, rugged sides and far-sweeping flanks — all this impressed me so indescribably, enthused me so thoroughly, that I then and there vowed, almost with fervency, that I would some day stand upon its glorious summit, if that feat were possible to human effort and endurance.[6]

Stevens and Van Trump met that same fall, and, from their common interest in Mount Rainier, they agreed to try an ascent. The summers of 1868 and 1869 were not favorable for the attempt as the smoke from settlers' land-clearing fires hung in a thick pall over the Puget Sound country. Their hopes might never have materialized if Stevens had not met a gentleman who sparked the venture.

Edmund Thomas Coleman was an "original member" of the British Alpine Club, taking part in expeditions in the Swiss Alps between 1855 and 1858.[7] His ascents of many of the principal peaks supplied material for the writing and illustrating by which he earned his living. In 1864 he took up residence at Victoria, on Vancouver Island, soon finding an outlet for his mountain climbing interest in a series of attempts on Mount Baker. On August 17, 1868, he reached the summit of that peak, and so was ready to try another of the giants of the Cascades.

A chance meeting of Stevens and Coleman early in the summer of 1870 found common ground in their interest in mountain climbing. Stevens had a mountain in mind but lacked the know-how to climb it, while Coleman was looking for another peak to try his mettle and experience. Stevens admits they expected the veteran English climber to be a great help on the upper slopes of Mount Rainier, but little did he know what differences lay between them, in habit, thought, and action.[8] Of their "guide, philosopher, and friend," Van Trump says: "Unfortunately for us in the matter of rapid progress and more unfortunately still for himself, our new acquaintance proved a clog on the expedition rather than an aid to it."[9] But for the moment they were happy with their new comrade, and plans for the grand assault were laid.

Subsequent events come into clear focus with a brief notice in the Olympia *Transcript* of August 6, 1870, which reads:

> Ascent of Mt. Rainier—E. T. Coleman, Esq., of Victoria, is in town preparing for a trip to Mt. Rainier, and will leave in a day or two. He will be accompanied by Gen. Hazard Stevens and Mr. P. B. Van Trump, of this place. The party expects some difficulty in reaching the top of the mountain, but will go prepared for the trip, and will accomplish the task if it is possible to do it.

The day before, Stevens and Van Trump had driven out to Yelm Prairie to solicit the advice and help of James Longmire, whom we have already met. Though they made good time over the hard dirt road, it was night before the prairie was reached. From it, the mountain they proposed to climb appeared "startlingly near and distinct" in the bright moonlight, so that their admiration was mingled with dread.

The night was spent at Longmire's big farmhouse and, at the breakfast table, the purpose of the visit was broached. Their host was adamant that a guide was essential if they were to find the route up the Nisqually River, reminding them the old Packwood trail had not been traveled for four years. Though there was no one else who could help them, he was reluctant to leave his harvest and his cattle business for so long, to which was added all the discouragement his wife could manage. A compromise was finally reached, whereby Longmire agreed to act as their guide to the end of the trail, then find an Indian to help them, after which he would return home.

With that settled, Stevens and Van Trump drove back to Olympia where they met Coleman and spent an evening examining the equipment he had brought with him for use on the trip. His gear covered the floor of the hotel room; every item of it considered indispensible, according to Stevens. There was a ground sheet, a rope, a pair of "creepers" (very similar to the crampons now in use), an ice axe, a spirit lamp, green goggles, deer's fat for the face, an

33

alpenstock, several remedies, a plant press, sketching materials, and some small items. Overall, Coleman's outfit would meet with the approval of present mountaineers, but his comrades in that venture were frankly skeptical of it.

Whether or not Stevens and Van Trump made any attempt to similarly equip themselves is unknown, but they did add an item not on the Briton's list. The amusing little story behind it was fortunately preserved in the reminiscence of one of the young ladies who accompanied the climbers to Yelm Prairie, their point of departure.

Several weeks before the start, General Stevens had material for a flag all cut and ready to sew, and this he took to Fanny Yantis, one of the young ladies who was to be a member of the party. He asked her to make the flag as a favor to him. Miss Yantis put the bundle aside, entirely forgetting that the flag was not made. When packing her baggage on Saturday for the early Monday start, she discovered her oversight. She rushed to Lizzie Ferry, who was also going on the trip, and begged her to help her out. Their principles forbade sewing on Sunday. So they laid out the pieces, threaded needles, and sat waiting for twelve o'clock to strike. The minute it was midnight, they went frantically to work. By six o'clock Monday morning, just before the start, the flag was finished. General Stevens never knew of his narrow escape.[10]

That hastily-made flag was certainly unique. Lacking time to sew the proper number of stars into the union, the girls remembered the original flag with its thirteen stars — and used thirteen also. However, the stars were placed in rows instead of in a circle![11] Odd as it was, Stevens had his flag; no, he had two flags, for he borrowed another from W. H. Cushman. The Cushman flag was also unique, for it had been brought across the plains to California during the gold rush, and had only thirty-two stars in its union, instead of the thirty-seven that would have been proper in 1870.[12]

On Monday morning, the eighth of August, half a dozen carriages, filled with young ladies and gentlemen, swept gaily through the tree-lined streets of Olympia, past the white, vine-covered houses so reminiscent of New England, to escort the mountaineers to the edge of the settlements, thirty-two miles eastward. After a short mid-day rest at

34

James Longmire's place, the party continued on to Lacamas Prairie, to the house of his son, Elcain. There a picnic supper was spread on a beautiful, wooded knoll nearby, after which the ladies retired to the roomy log house for the night, while the men unrolled their blankets on the ground under the trees.

But there was little rest for the weary in that encampment. Coleman says that he and his comrades had no sooner lain down than their boisterous friends began a "double-shuffle" over them. The night was made hideous with war whoops and animal calls, and the horseplay became very rough indeed.

James Longmire arrived early the next morning bringing a horse and two mules, one of which he was riding.[13] The pack animals were soon loaded, and after farewells which included salutes from the available firearms, the carriages took the road to Olympia, leaving the little party of mountaineers to follow their guide eastward into a great forest known but imperfectly even to him.

After leaving the Lacamas Prairie, the route passed over timbered, hilly ground for three miles, then descended into the Nisqually River bottoms where it wandered for ten difficult miles through the dry channels and vine maple jungles south of the river. The unaccustomed exertions of the three footmen among the sultry thickets and mosquito infested sloughs distressed them so much they probably doubted the wisdom of their plan to walk in order to get into condition for the mountain climbing ahead. Longmire had his troubles following those scattered, partly overgrown blazes, losing several hours in searching for the trail, but he finally brought them to a ford of the river just above the mouth of Ohop Creek.

While crossing the river, an incident occurred which came near ending the expedition. At that point the stream was a waist-deep, icy flood of milky colored water one hundred yards wide; so the animals were used to get over. Van Trump led on the saddle mule, with Coleman mounted behind him holding the neck-rope of the pack mule, which carried Longmire as an additional load. He, in turn, held the neck-rope of the packhorse on which Stevens was mounted. Soon after entering the water, Coleman let go of his rope and the two pack animals turned down stream, requiring strenuous effort on Longmire's part to keep them from the rapids below the ford. That was the first of a series of events which soon brought Stevens and Van Trump to regard their English comrade as primarily a hindrance.

With the river safely crossed, a zigzag trail took the party up the west end of the high, sandy bluff opposite the ford. After gaining the top, the route followed the brink above the river for a mile, then turned through the forest to come out on Mashel Prairie, where they camped for the night near a spring on the edge of that beautifully embowered meadow.

All were weary from the hard day's travel over rough country, yet the chores of camp life had to be done; the stock had to be picketed in the meadow, there was firewood to cut, cooking to do and beds to be made, and it did not improve Coleman's popularity when he sat down with his pipe and notebook, to smoke and sketch while the others did the work (though he did determine the elevation of the prairie as 800 feet above sea level, by the barometer). Despite a growing disappointment in their climbing companion, Stevens and Van Trump were in high spirits at that encampment, for the two flags fluttering so gaily from their alpenstocks "made failure seem impossible."

The following morning, August 10th, they were on the trail early, though Van Trump hints that Coleman was no help in that either. Another of his peculiarities was a stubborn insistence on two baths a day — a plunge in a convenient stream or a sponge bath was an absolute necessity "twice a day, with all the regularity and certainty with which that division of time begins and ends." Those ablutions were undoubtedly a comfort and of benefit to Coleman, but his traveling companions saw in them only another excuse for failure to accomplish a fair share of the camp work, and therefore, a cause for delay.

As the party left Mashel Prairie, Longmire pointed out the oak tree where two Indians were hanged by the territorial volunteers under Captain Maxon during the Indian war of 1855-56. However, the battle which took place a mile from there, on the Mashel River, was not as bloody as he represented it.[14]

Entering the forest again, the route led to a large "burn," beyond which were some deserted Indian huts where the trail they were following joined another from Nisqually Plains.[15] The route then descended through swampy timber to the clear Mashel River near the present town of Eatonville. There the animals were unsaddled for a brief noon rest, and while Longmire crossed the stream to scout out the trail, Van Trump went fishing and Coleman enjoyed a solitary tea.

An hour later Longmire returned with the disappointing news that he had lost the route in the tangle beyond the river. The party was soon in motion to cross the Mashel, but not without a ducking for the guide, whose saddle mule went down while trying to scramble up the far bank. Dripping wet, he led off.

They passed through thickets of ferns, salal and Devil's Club, over logs and around the prostrate trunks of huge fallen trees. There in the mossy gloom of the old forest, men and animals made slow, crashing progress, with frequent halts to adjust loosened packs. When almost up the hill east of the Mashel, the packhorse fell under his unbalanced load and rolled down, so that it was necessary to unpack him and carry the goods up piecemeal.

Once upon the top, the route was better, as they traveled south through forest and marshy swale to a camp on the bank of the Nisqually River, near the ruined log house of a former Indian medicine man who had paid with his life for his failure to cure a patient.[16]

The route of travel on August 11 lay up the north bank of the Nisqually through vine maple thickets and over fallen timber to Silver Creek, where a noon stop of an hour was made.[17] Beyond the creek, the Nisqually Valley opened to them as they entered the old burn which was a landmark for Kautz and may have been enlarged by his berry-picking fire! The fine view of the mountains and Mount Rainier, now so startlingly near, did not ease their difficult route through the down timber with which the burn was strewn.

At dusk a third camp was made at Copper Creek, after thirteen hours of fatiguing travel.

Daylight found the party moving again. From the mouth of Goat Creek they were obliged to follow the gravel bars along the river for a time, then they traversed the hillside from the present park entrance to Tahoma Creek, which Stevens called the "Takhoma branch or North Fork." After crossing that brawling thirty-yard stream, white with glacial silt from the southwest slope of Mount Rainier, a way was found directly up the bed of the Nisqually by crossing and recrossing between gravel bars.

It was slow going, requiring five hours for the three miles from Tahoma Creek to the point where they turned away from the river toward Bear Prairie. At the last crossing of the river, there was more of the misfortune that had earlier been their lot. The packhorse went down in a slough, making it necessary to remove the load to get the animal up. While unloading, the flour sack was found to have a rent in it and had to be sewed where they stood, knee-deep in the cold water. Coleman's comment: "Never was sewing performed before under greater difficulties."

From that crossing, their route led up Horse Creek, then over a low, fire-swept divide to Bear Prairie, where they encamped beside a rivulet near the east end. The prairie was a narrow, green oasis surrounded by the charred ruin of a forest (a desolation which extended eastward down Skate Creek to the Cowlitz Valley and down the Cowlitz Valley to the land-clearing fires of the upper settlements). From there had come much of the smoke blanketing Mount Rainier during the summers of 1868 and 1869. The fastidious Coleman named that camp amid the ashes "Camp Dirty," which Van Trump countered with a remark that it would give them grit for the climb to the summit.

Van Trump did not feel kindly toward the Englishman, for he had come to consider him a mere drag upon the expedition. He was particularly disgusted with one of Coleman's exaggerated habits on the march — an insistence on drinking at each little stream crossed. The act of drinking was a careful ritual, in which a folding cup was used to pour stream water into a partly-filled brandy flask, after which a drink was poured from the mingled waters. This indulgence often left Coleman behind with the result that he frequently became lost and the party was delayed searching for him. All the arguments Stevens and Van Trump could find in support of a swift march were met with the remark, "we didn't travel so in the Alps." For his part, Coleman thought his companions were out to "do" the mountain, which was as true as it was repugnant to the Briton!

As the weary mountaineers relaxed around the fire that night, tragedy again came near to putting an end to their hopes for a successful ascent. One of the fire-blackened trees standing near the camp began to creak. It was a barely adequate warning followed by the crash of the great trunk across a pair of recently occupied blankets. Who the lucky one was, is not recorded.

On the following day, August 13, Longmire and Stevens went down Skate Creek to its junction with the Cowlitz Valley to locate Poniah's band of upper Cowlitz Indians from whom he had hoped to obtain a guide.[18] Though the toilsome journey brought them to a deserted camp, their effort was not wasted because of the chance finding of a single family living in a rude shelter nearby while awaiting the ripening of the berries.

It was the lodge of that same Sluiskin who fled wounded from the battle of Grande Ronde. Rather than submit to reservation life, he had slipped away to the mountains, appearing at the camp of Poniah with hawk bells and finery to pass himself off as a chief. A marriage to one of the headman's daughters was easily arranged and Sluiskin began the isolated existence in the upper Cowlitz and Cispus valleys which, thereafter, was his way of life.

He was a friendly fellow, this Sluiskin, welcoming Longmire and Stevens with dignified hospitality. With much hand shaking, they were seated beneath the little shelter of hides to be served cakes of dried huckleberries by his squaw; then there was talk through the medium of Chinook jargon. Longmire soon explained the purpose of the visit, obtaining Sluiskin's services as a guide, with the understanding that he would present himself at their camp the following day. Late in the evening, Longmire and Stevens returned to the camp on Bear Prairie, weary but with good news.

The next day was begun with industrious preparations for the assault on Mount Rainier. The mountaineers sorted equipment and supplies, put spikes in their shoes and fought off the mosquitoes and gnats that paid so little attention to their smudge fires. About noon Sluiskin arrived, true to his word. His entry was impressive; he rode up on a scrubby Indian pony with his young son behind him, while his squaw followed afoot carrying an infant on her back. And he was dressed for the occasion. A blanket, belted at the waist into a loose cloak, partly covered a buckskin shirt and a breechclout of striped woolen stuff, while buckskin leggings (in lieu of pants) filled in the gap down to his moccasins. But Sluiskin's headgear was the really distinctive part of his costume. It was made from an old army forage cap with the small, round crown replaced by the perforated brass base which once had held the chimney of a coal-oil lamp. Brass nails had been driven through the visor until they projected about an inch below, creating a spiny barrier before his eyes; the whole crested with eagle feathers. Stevens called the effect fierce and martial.

The new guide sat himself down to devour an ample dinner, while Longmire began his homeward journey, taking the mules with him and leaving the climbers only the little packhorse for the transportation of their camp gear. Sluiskin's bargain with the white men called for him to guide them to the base of Mount Rainier for a wage of one dollar per day, with his squaw receiving an allowance of rations for looking after the camp and the horse.

Now Sluiskin was a sharp fellow and he saw it would be to his advantage to prolong the trip — more days, more dollars! So he quickly vetoed Stevens' suggestion that they make the approach by way of the Nisqually River, insisting in his fluent Chinook, reinforced with pantomime, that the only practicable route lay along the summit of the Tatoosh Range, northeasterly from camp. However, he made it plain that he did not believe they could climb the mountain and he obviously considered the venture highly ridiculous.

In their ignorance of the locality, the climbers agreed to Sluiskin's route and he led off, his pack supported by a tumpline across his forehead and his long-barrelled Hudson's Bay trade gun held crosswise on his head. All afternoon they toiled upward in single file, Coleman lagging in the rear, often calling to the others to wait for him. He could not keep up, though his pack of between thirty and forty pounds was no heavier than the others, and by the time Sluiskin and his two hardier patrons passed onto the top of the long ridge which extends westward from Lookout Mountain, the Englishman was out of sight and sound.

As Coleman approached the top, he worked himself into a *cul de sac* among the little cliffs in which the ridge terminates.[19] Soon he was stuck, finding it necessary to jettison his pack which rolled and bounded down the rocks to disappear among the bushes below. Unencumbered, he was able to climb down, but he could not find the pack, and, without equipment, could only return to the camp on Bear Prairie.

On the ridge top, Stevens, Van Trump, and the guide had thrown off their packs for a rest while they waited for Coleman to come up. When he failed to appear, and their hailing went unanswered, Sluiskin was sent down to assist the missing man. He returned an hour later, to report that he had finally sighted Coleman near the foot of the mountain, trudging toward camp without his pack. Sluiskin was disgusted. Stevens and Van Trump were more concerned about the loss of their entire supply of bacon, which had been in Coleman's pack. But the loss was softened, no doubt, by their relief at being rid of so vexing an impediment as the Englishman, who would be safe at camp, while Sluiskin's gun would supply them with meat; the decision to push on was easily made!

The guide led them two miles along the south side of the ridge to a sheltered hollow where he had often encamped. However, the spring had dried up, so that Stevens had to go back a mile with a canteen and coffee pot to get water. It was a bleak, mosquito-plagued, and chilly encampment which they left at dawn breakfastless and thirsty.

From Lookout Mountain, they traveled northward along

the divide between the drainages of Horse and Johnson creeks, coming after two hours, to a snow bordered lakelet where they breakfasted on bread and coffee while shivering with the early morning chill. Soon on the move again, Sluiskin led them up a difficult climb onto the peak which now bears the name of Wahpenayo, a father-in-law of Indian Henry, to a spectacular view of Mount Rainier — a view so inspiring that Stevens says, "we were already well repaid for all our toil." A few more hours brought them to the saddle between Plummer and Pinnacle peaks where they gladly descended the steep, snow-filled gulch to the little stream they called "Clear Creek."[20]

There they halted for an hour and Van Trump couldn't resist the temptation to try a fly on its waters, though without success. Sluiskin was reluctant to go on, for he seemed worn out, but Stevens and Van Trump insisted in pushing on up Mazama Ridge. As they proceeded, the forest opened, allowing them to see Reflection Lake and note that it lay upon the divide between the Nisqually and Cowlitz rivers. Night overtook them among the scattered groves, where they found a camping place in a clump of alpine firs.

For the evening meal, the ration of dry bread and coffee was supplemented with a grouse shot by the guide along the way. Sluiskin's attitude toward his companions had changed greatly during the day's march. The derisive jesting so prominent after Coleman's defection, softened to a genuine concern for them, and that evening, as they sat by the campfire, he attempted to dissuade them from trying the climb to the summit.[21]

In his Chinook and broken English, he told them of the difficulties and danger awaiting them. Here are his words, as well as Stevens could render them forty-five years later:

> Listen to me, my good friends. I must talk to you.
> Your plan to climb Takhoma is all foolishness. No one can do it and live. A mighty chief dwells upon the summit in a lake of fire. He brooks no intruders.
> Many years ago my grandfather, the greatest and bravest of all the Yakima, climbed nearly to the summit. There he caught sight of the fiery lake and the infernal demon coming to destroy him, and he fled down the mountain, glad to escape with his life. Where he failed no other Indian ever dared make the attempt.

At first the way is easy, the task seems light. The broad snow-fields over which I have hunted the mountain goat, offer an inviting path. But above them you will have to climb over steep rocks over-hanging deep gorges where a misstep would hurl you far down—down to certain death. You must creep over steep snow banks and cross deep crevasses where a mountain goat could hardly keep his footing. You must climb along steep cliffs where rocks are continually falling to crush you, or knock you off into the bottomless depths.

And if you should escape these perils and reach the great snowy dome, then a bitterly cold and furious tempest will sweep you off into space like a withered leaf. But if by some miracle you should survive all these perils the mighty demon of Tak-homa will surely kill you and throw you into the fiery lake.

Don't you go!

You make my heart sick when you talk of climbing Takhoma. You will perish if you try to climb Takhoma. You will perish and your people will blame me.

Don't go!

Don't go!

If you will go, I will wait here two days, and then go to Olympia and tell your people that you have perished on Tak-homa. Give me a paper to them to let them know that I am not to blame for your death.

My talk is ended.[22]

Long after the two climbers had wrapped their blankets about them, Sluiskin sat by the fire chanting dismally, accompanied by the sound of distant cataracts and the occasional muted rumble of an avalanche on the formless white slopes, standing so coldly beautiful above them that moonlight night.

On the following morning, the party moved up the ridge to the timber line where a base camp was established in a clump of alpine firs on a knoll near Sluiskin Falls, which they named for their guide. The stream which provided the falls, they called Glacier Creek,[23] and the ice which was its source (so close then it seemed to menace their encampment), they designated the Little Nisqually Glacier.

After lunch, Stevens and Van Trump set out to recon-noiter the mountain. They traveled up the steep, sloping sheet of ice extending from the terminus near their camp-site to the base of what is now known as Anvil Rock. The route led them easily onto the Muir snowfield, from which they ascended the Cowlitz Cleaver to a point having a good view of the western face of Gibraltar Rock, though it would not be known by that name until much later. Van Trump says, "after a long and careful study of it from our point of view on the Cleaver we felt pretty confident that on the morrow we would be able not only to effect the passing of the cliff, but also the ascent from there." And so, the two climbers returned to their encampment, cheerful and with high hopes for the morrow.

During the night, a freshening wind drove sparks from their warming fire into the scrubby, weather-beaten trees about the camp. There was a brilliant conflagration, harbinger of the havoc carelessly handled fire would later bring to the beautiful alpine groves.

On August 17, which was Wednesday, the climbers were up before dawn; even so, Sluiskin had already taken his gun and left. By six o'clock they were off to the summit, convinced that they could make the round trip in one day. In addition to alpenstock, creepers, rope, and ice axe, they carried a canteen, some lunch, gloves, green goggles, a brass plate, and the two flags, but no coats or blankets. Omission of the latter they would regret.

Several thousand feet above the encampment the climbers saw Sluiskin, sitting motionless upon a high rock to the right of their line of ascent. He was facing the mountain, studying it intently, and his only response to their hail was a glance. Progress was so rapid Stevens and Van Trump required only three hours to reach the point on the Cleaver to which they had ascended the previous day.

Above that point the climbers kept upon the top of the ridge as much as possible, occasionally passing rocky obstructions by dropping down onto the flanks of the ridge. In that manner they progressed to the shallow, wind-swept saddle which is now known as Camp Misery, then quickly found the debris-covered, sloping shelf of the "ledge," that dangerous passage around the western side of Gibraltar Rock. There they crossed on treacherous footing, a thousand feet above the crevassed and rock-laden surface of the Nisqually Glacier, under an intermittent bombardment of

44

small stones and debris — the same rock fall that has been the *bete noire* of climbers ever since.

The two climbers crossed safely to the steep chute formed at the juncture of the cliff and the glacial ice which descends in chaos from the dome of the mountain. There the danger of being struck by free falling stones, thawed loose from their frosty moorings, was augmented by the danger from rocks sliding and bounding down that icy groove which they had to ascend. Van Trump was soon struck by a small stone, while another knocked the staff from his hands, sending it sliding down to the Nisqually Glacier far below. Van Trump then used the ice axe to cut steps up the ice to their left, and so got them out of the chute onto the safer ice extending upward to the summit.

The first axe to strike cold, stinging chips from the glacial ice of Mount Rainier is an interesting implement. It was Coleman's, copied from the primordial ice axe Whymper had shaped from a British naval boarding axe. Contemporaneous with the one that conquered the Matterhorn, Coleman's axe had also been carried through the Alps in the very dawn of mountaineering, then to a "first" on Mount Baker before it was carried to the summit of Mount Rainier. Now it lies somewhere in the large summit crater, lost there since 1892, but that part of the story is best saved until later.

Once up out of the chute, the ice was rough enough to afford good footing and the climbers moved rapidly up to where the dome was less steep. They continued to bear to the left, traveling diagonally upward over the sun-cupped surface, crossing several crevasses without difficulty at narrow places before they came to a great one. It was from eight to twenty feet wide, presenting its greenish depths as an unbroken barrier to farther progress.[24] Fortunately, the upper wall of the crevasse was considerably higher and tilted in a manner to overhang the lower lip in places. At such a place the climbers managed to cross by throwing the bight of their rope around an ice pinnacle and ascending hand-over-hand for twelve feet.

Above the great crevasse Stevens and Van Trump began to feel the wind, which soon became a chilling gale, from which they found a slight protection by wrapping their bodies with the flags they carried. They were also troubled by fatigue brought on by their exertions and the rarity of the atmosphere, so that the rate of travel became slower, with frequent stops to rest. Stevens says, "after taking seventy or eighty steps, our breath would be gone, our muscles grew tired and strained, and we experienced all the sensations of extreme fatigue. An instant's pause, however, was sufficient to recover strength and breath."

The course followed brought the climbers to the southwest peak, which they found to be a narrow ridge projecting from the summit dome. Meanwhile the wind had become so violent that the last fifty yards along the ridge was traversed on hands and knees, sheltering where possible behind ice hummocks and in fissures in the snow. At last they were able to stand upon the summit of what they called Peak Success, where they raised their flags on alpenstocks, to pound and snap, while they gave three cheers that were almost soundless in the bitter blast.

Standing there as they were, on a freezing, windswept point from which the snow fell away in an ever steeper slope on the one hand and bare rocks dropped away into mists on the other, the climbers became aware of their perilous position. It was too late to descend safely by the route they had come up, yet there was no shelter where they stood that would serve them, chilled, fatigued, and insufficiently clothed as they were. They might have dropped down to some loose rock below them, to burrow for shelter there, but they didn't. Instead, the decision was made to push on to the true summit, several hundred feet above them in elevation.

It was a fortunate decision. Surmounting the rocks which rim the main peak, Stevens and Van Trump found themselves within a snow-filled crater, 200 yards in diameter, where steam issued from the rocks along the northern side. After warming at a jet, they looked about for a place to spend the night, soon finding shelter in a cavern hollowed out by the heat and steam between the ice and the sloping wall of the crater.

Forty feet from the entrance, the climbers built a wall

of stones around a strong steam jet, giving them a space five by six feet where they were secure from the gale, except when violent gusts were deflected downward by the sloping roof of green ice overhead. In that cramped shelter they ate their lunch and then sat through the hours of darkness, blistered by the hot steam on one side, frozen on the other, and nauseated by the sulphurous odors. It was a miserable night of fitful dozing, but no real sleep.

In the gray light of dawn, the continuing gale drove a thick mist over the summit, obscuring all but near objects and raising the fear that a developing storm might trap them there without food. Nine o'clock brought occasional rifts, so preparations for the descent were quickly made. Before starting down, Stevens made a brief sally to what he called, "a huge mount of rocks on the east side of our crater of refuge, which we named Crater Peak." There he left an inscribed brass plate, from which they had deleted Coleman's name with the point of a knife, before placing it in the cleft of a boulder with their canteen.[25] And so, that morning of August 18, 1870, a certain conquest of Mount Rainier was formally completed.

Stevens retreated in a nearly frozen condition to the cavern, where both climbers exercised violently to restore circulation, before leaving their boreal shelter. As they passed over the icy knoll which forms the highest summit, another and larger crater was found lying to the east. They were also allowed a tantalizing view of the North Peak through rifts in the mist and clouds gathering in a "weather cap" over the summit; though the temptation to visit it was great, they dared not delay longer than to bestow the name "Peak Takhoma."

The descent was begun without difficulty. The great crevasse was crossed on a snowbridge which had been overlooked on the ascent, while the steep chute was descended by fastening the rope to a rock to serve as a hand line, though they had to abandon it there. Four and a half hours after leaving the summit, the dangerous "ledge" was behind them, and they proceeded to recover the alpenstock Van Trump lost on the ascent.[26]

The remaining distance to the encampment at Sluiskin

47

Falls should have been innocent of danger for such veterans, but it was not so. When nearly to camp, Van Trump suffered a painful injury in a fall on a small snowbank. One of his clumsy creepers turned under his foot, letting him slide forty feet onto the jagged rocks below. Fortunately, his injuries were no worse than abrasions on the hands and face, bruises and a deep gash in the thigh.

Stevens helped his injured comrade into camp and then set about getting dinner. Coffee was soon boiled and some roasted marmot readied, for that was all the meat Sluiskin had managed to get during their absence. The little animals were tough and strong, which caused Van Trump to call their *piece de resistance* "raw dog," and only their great hunger brought them to eat it.

As they were finishing, Sluiskin was seen approaching, weary from a fruitless hunt for goats. When he saw the climbers had returned, he stopped and considered them warily for a time before coming into camp. Sluiskin was sure the two men by the fire were ghosts, for hadn't they gone to certain death on the mountain? Finally convinced they were really alive, he was as generous with his praise as he had been scornful before. They were "skookum tilicum, skookum tumtum"—strong men with strong hearts.

The next morning a start was made toward the camp on Bear Prairie, but Van Trump was able to go no farther than Clear Creek. There it was decided to leave him encamped while Stevens and the guide went back to the prairie for a horse. Accordingly, the injured climber was made comfortable and left with all the remaining food (enough to last him a week), at a camp which would be marked for many years by a rusty ice axe stuck firmly in the trunk of a tree.

Stevens was determined to find a better route than the one over the Tatoosh Range; accordingly he started down the valley toward the Paradise River despite the protests of Sluiskin. Soon the guide attempted to veer off to his former route and Stevens had to take the lead in a rather blunt manner. Upon reaching the Nisqually River, at the mouth of the Paradise, Sluiskin became cooperative, cutting the time from there to Bear Prairie by half through his knowledge of the country.

˙ Coleman was found safe at camp, leisurely preparing a one-man assault on the summit. He had some cricketer's spikes for his shoes, and had made two caches of food on the mountain top above Bear Prairie, but the return of Stevens and the guide put an end to his plans.

The next morning Sluiskin left with two horses and his little boy to bring back the injured Van Trump. Meanwhile Coleman retrieved the food caches he had made and Stevens moved the camp over to the Nisqually River, which would save the injured man some painful travel. There he made a shelter from Coleman's gum sheet with a roaring fire in front to offset the drizzling rain.

Sluiskin managed to get the horses as far as the Paradise River by following up the bed of the Nisqually. He then left the horses with his boy and went to get Van Trump, whom he assisted through the forest and over the rock slides with the utmost kindness and concern. While they were passing the spot where he had attempted to mislead Stevens, Sluiskin told with evident good humor how the "skookum tenas man wah-wah hi-u goddam"— so we may presume some strong words were spoken on that occasion!

Late that afternoon the relief party reached the new camp on the Nisqually. Sluiskin then took the horses back to Bear Prairie for grass while Van Trump was made as comfortable under the gum sheet as the weather permitted. The storm continued, and on the second day, Sluiskin returned the packhorse, saying he was moving to the Cowlitz with his family, as he could not feed them at Bear Prairie. The Indian felt bad about his failure to kill a goat for them while on the mountain, promising to bring both Stevens and Van Trump a skin the next summer, which he did. For Coleman, he had only some unkind words in Chinook and broken English; it is as well the Englishman did not understand.

The party remained under their scanty shelter, in sodden dampness, until August 24, which dawned clear, allowing them to start the return trip. They found the Nisqually River swollen from the rain, and the frequent wading of

49

its turbulent waters satisfied even Coleman, who remarked "the party had no lack of cold bathing."

That night they reached Silver Creek, and the next, despite difficulties in finding the trail, they were able to stop near the ruined house of the unfortunate medicine man. The packhorse was unable to find enough forage, and gave out during what they hoped would be the last day on the trail. It was necessary to jettison much of his load, a delay which forced them to encamp after dark, less than a mile from Elcaine Longmire's farm.

After a short rest at that homestead next morning, they went on to James Longmire's place on Yelm Prairie, where a triumphal feast awaited them. From there, they were returned to Olympia in Longmire's family carryall; and so, with the two flags flying from alpenstocks, fastened one on each side of the vehicle, they entered town, as Coleman says, "literally with flying colours amid general congratulations."

Van Trump was more than two months recovering from his injuries, but the time must have passed pleasantly enough, for both he and Stevens were the heroes of the hour, though of course, there were the scoffers who held a forum beneath the great maple tree across from Mrs. Steele's hotel; their wisecracks left lasting scars on both climbers. The unlucky Coleman slipped off to obscurity in his writing and illustrating, and it must be noted that all he wrote about the expedition was scrupulously fair to his two companions, and entirely lacking in the rancor with which they presented him. Coleman was a thorough gentleman, if not a determined mountaineer.

And what of Sluiskin-of-the-broken-hand, the independent, jocular, wily, yet entirely loyal guide? He lived on in his favorite mountains and valleys, prospering a little but changing not at all. Father P. F. Hylebos, of the Cowlitz Mission, knew Sluiskin in his later days, once undertaking to lecture him on the evils of polygamy, for the old Indian had accumulated three wives by then. The good priest told him, "The great white father in Washington doesn't like a man who has three wives." Sluiskin smilingly replied that he found three quite comfortable; in the morning there was one to put on his right moccasin, one to put on his left, and one to prepare his breakfast.[27] That was Sluiskin, and he, no doubt, savored life to the day he drowned in the Yakima River, near the turn of the century.

This account of the most important ascent of Mount Rainier properly closes in the office of the editor of the Olympia *Transcript* where Stevens and Van Trump were being interviewed shortly after their return from Mount Rainier. Captain Tyrrell, who was considered a "half-crazy old character," was present and interested.[28] After Stevens told of climbing over ten miles of rock and snow above their camp, the old fellow laughed heartily, and said: "Why, when I went up that mountain I traveled over thirty-five miles of snow, and instead of an alpenstock I used a spur of the mountain to climb with."[29] I can almost hear the guffaws! But the real joke is that his "spurious" ascent has since been given some credence.[30]

While Stevens and Van Trump were making the first indubitable conquest of Mount Rainier, the United States Coast Survey was taking its measure in the truest sense — by determining the elevation of its summit. The work was accomplished by James Smyth Lawson, who took vertical angle observations from a station fifty-eight and a half miles west of the highest point. Subsequent computations established the elevation as 14,444 feet (only thirty-four feet higher than the present elevation). However, the result "was considered to be uncertain by several meters," due to a lack of adequate checks on the field work, and it was not brought to the attention of the public until eighteen years later. In that, the Coast Survey was acting with the same spirit of caution it displayed in withholding the results of the 1856 observations.[31]

Eighteen hundred seventy is a year which cannot be dismissed without briefly considering another event, anti-climax though it is. Late summer found the work of the Geological Survey of the Fortieth Parallel, a sprawling inventory of natural resources along the routes of the Union Pacific and Central Pacific railroads, nearing completion at Mount Shasta in northern California; so Clarence King, who was directing it for the Corps of Engineers, was able to send two of his assistants northward to obtain data on volcanism, intending to follow them later. The two surveyors, Samuel Franklin Emmons and A. D. Wilson, arrived in Olympia when the excitement over the ascent of Mount Rainier was at its highest. Deciding immediately to ascend the mountain themselves, they called upon General Stevens, who gave them detailed information, before sending them on to James Longmire on Yelm Prairie.

But Longmire's services as a guide to Mount Rainier were not easily obtained. Emmons says:

> . . . he was firm in the opinion that it was too late in the season to undertake the trip with safety, and the opposition of his wife, who averred that the fatigues had more effect upon him than a month's sickness, was even more pronounced. I really believe that it was only a feeling of duty for the advancement of science, and a personal regard for us, fearing that some misfortune might befall us should we undertake the trip alone . . . that finally decided him.[32]

And so, about noon on September 27, the party of four men and five animals set out to retrace that tortuous route up the Nisqually Valley. A number of mules and a quantity of provisions were left behind to be brought on by Mr. King.

Emmons and Wilson had been working for three years in the vast, open reaches of the plains and deserts, and the effect produced upon them by the great forest into which they plunged was marked.

> Our first few miles were exhilirating in the extreme . . . we passed through endless aisles of stately pines, cedars and firs, whose size increased as we progressed . . . the solemn silence, scarcely broken by the dim rustling of the wind on the distant summits of the trees, and the semi-twilight which the density of the foliage overhead produced, at first quietly refreshing, became ere long gloomy and oppressive.

Nor were they cheered by the accidental breaking of a cistern barometer, making it necessary for them to send their campman back with one of the animals.

The two surveyors were properly appreciative of the woodcraft of their guide, but they discovered at the first night camp, on Mashel Prairie, that he was "innocent of the art of cooking." That old pioneer, Indian fighter and explorer? Hardly! The boys also got the job of packing the

52

mules, for their guide claimed he couldn't understand their Mexican method.

About noon on the fourth day, the persistent rain stopped and a partial clearing gave them a view of Mount Rainier. They camped that night where the Indian trail crossed the Nisqually below Bear Prairie Point, which was as far as Longmire had been up the river.

The following morning they made their way up the bed of the Nisqually to its junction with the Paradise River, hoping to ascend that stream by the route Stevens and Van Trump had used returning from the encampment in Paradise Valley. They managed well enough for a short distance, but soon became entangled in the brush-covered rock slides where the canyon narrows between Eagle Peak and Ricksecker Point. In the struggle to get the pack mules through, both animals went down and were only extricated after the loss of the remaining barometer. Abandoning the route, they returned to the Nisqually to make a night camp.

Emmons and Wilson went afoot up the Nisqually from that camp, satisfying themselves there was no chance to get the animals up the steep, heavily timbered east wall of the canyon. So, after exploring about the snout of the glacier, they rejoined Longmire and moved back down the river to what they facetiously called "Longmire's trail."[33] Two days' travel along that primitive, fire-blackened trace brought them into the Cowlitz Valley.

There they found a group of Indians living in lean-tos made of cedar slabs, and obtained two guides at the old rate of a dollar per day. They were then conducted along Backbone Ridge to the edge of the snowfields, where a campsite was found near a clump of alpine firs overlooking the Cowlitz Glacier, 1,500 feet below.

An assault on the mountain was deferred in expectation of the arrival of Mr. King with another barometer, the two surveyors occupying themselves with exploring the lower slopes. A trip was made "to the northern flanks of the mountain, beyond White River," carrying blankets, provisions, and their instruments with them. A few more days were spent in exploring about the heads of the Cowlitz, Nisqually, and "Och-hanna-pi-cosh" rivers, and in measuring a base line from which they could begin the triangulation for a map. Longmire soon had enough of scrambling about on the glaciers, contenting himself with caring for the camp until the weather took a threatening turn, whereupon he began the return trip to Yelm Prairie. The one Indian who had stayed with them (Muck-a-muck, the hungry) also left, leaving Emmons and Wilson to the risky business of outwaiting a fall storm.

They spent four days sheltering from a southwest gale on a daily ration of one cup of flour and two spoons of coffee each. When the weather fortunately cleared, they quickly prepared for an ascent of the mountain, for they feared it might be the last chance. Three cups of flour each were baked for three days' provisions, blanket rolls were made up and they were ready.

After crossing the Cowlitz, a high camp was found on the west side near the dead trunk of a gnarled pine, the highest firewood available. The remainder of the day went into reconnoitering the upper slopes from nearby points; as a result, a one-day ascent was decided upon, with a start at three o'clock the following morning. But Wilson suffered from neuralgia during the night and they did not get off until 4 a.m.

The moon favored them at the start, giving way to full daylight before they reached the foot of the Cowlitz Cleaver three hours later. That far, the route was an easy one across ice broken by rocky outcroppings, but, once upon the great ridge of burnt rock which slopes steeply up to a junction with the bluff face of Gibraltar Rock, the climbing became slower — and more dangerous. The great blocks and loose scree provided poor footing, while the exposures were often nerve wracking, yet two hours of hand and foot climbing sufficed to bring them under the great thousand-foot cliff.

They passed carefully across the west face on a narrow ledge, clinging tightly to the wall on their right in an effort to avoid the rocks which fell singly and in showers upon the outer edge, to rebound toward the Nisqually Glacier far below. Where the ledge ended, at the junction of the cliff and the cascading ice of the glacier, they found the gutter from which Stevens and Van Trump had been driven by the rolling, bounding rocks. It was steep, winding, and just wide enough for a man's body, and in it, they found a heavy rope hanging.

It was the rope left by their predecessors two months earlier. Emmons approached it cautiously, giving it a tug in test. But the rope pulled free; so he coiled it about his waist for future use. Taking his geologist's hammer from his belt, he advanced into the chute to hack steps. As he worked, the edge of his pack (a knapsack with an overcoat roll upon it) struck the projecting wall, disturbing his balance, so that he dropped the pack. The two climbers had to stand and watch it cartwheel down the butter and over the brink, powerless to stop it. Gone was their brandy, coffee, and firewood, and worst of all, Emmons' overcoat and the ice creepers made for him by an Olympia blacksmith.

The climb was continued with great caution. Emmons went up by means of toe notches to a ledge forty feet above, hauled up Wilson's instrument on the rope, with him following, then cut steps to a higher perch where he again assisted the heavily burdened topographer. All the way up the chute they were under a rain of pebbles, but around one of its windings, they came upon a real terror—a large boulder nearly melted out and poised threateningly above them. There was no alternative to passing it, yet they did so with misgivings.

From there on the going was better. A short scramble over roughened ice brought them to a large crevasse, which they were able to cross on a snowbridge, using their rope "after the most approved Alpine fashion."[34] Two more hours over an icy slope of thirty degrees, with no crevasses larger than they could jump, brought the climbers to the summit at one o'clock, October 17, 1870.

Deprived of the warming exercise of the climb, Emmons began to suffer from the violent gale, "which seemed almost sufficient to blow away the very rocks." Lacking his overcoat, the wind and the 32° temperature soon robbed him of his body heat, forcing him to retreat into an ice cave to thaw frostbitten fingers at a steam jet. Wilson attempted to set up his theodolite upon the fifty-foot hummock of ice which is the highest point, but the force of the wind swept both him and his instrument from its top, depriving him of his opportunity to establish the height of the summit by an observation of the depression angle of the sea horizon.[35]

Two hours were all they could spare upon the summit, and the return was begun at three o'clock. From the northeast rim of the crater, they looked down the long cascade of ice leading to White River, observing how that greatest of Mount Rainier's great glaciers (which would finally be given Emmons' name) was so steep on its upper reaches the view was like looking over a wall.

Without his creepers, Emmons found the descent of the dome most perilous. Using the spiked tripod of the useless theodolite to scotch his footholds, and with Wilson ahead as security against an inadvertent glissade, the dangerous passage was made. At the chute, the boulder which tested their fortitude on the upward climb was gone, as were most of their steps, but they descended easily by means of the rope, which they left firmly anchored there.

Darkness caught them at the foot of the Cowlitz Cleaver, leaving them to pick a difficult way to camp by starlight. The elapsed time was seventeen hours for that first one-day ascent by the Gibraltar route.

Unable to wait longer for Mr. King,[36] for lack of provisions, and doubtful of their ability to follow the dim trail to Yelm Prairie, Emmons and Wilson abandoned their base camp and made their way across the Cascades to the Indian Agency at Fort Simcoe, to find they had been given up for lost. They were famished and in tatters, but had accomplished much.

In due time, Emmons wrote to his chief, giving a good description of Mount Rainier from a geological viewpoint. Clarence King thought so well of the letter that he had it published, thereby involving Emmons in an unfortunate controversy with Hazard Stevens.[37] The published letter provided a very poor description of the climb made by Emmons and Wilson, from which Stevens reasoned they had not reached the top. The dispute was publicly aired without accomplishing any good purpose other than to bring into print the first map of the upper reaches of Mount Rainier.[38]

In 1877 Emmons presented before the American Geographical Society, a paper which acquits him of all imputation that he did not reach the summit.[39] There can be no doubt that he stood upon that gale-swept peak, fixing in his mind the geological features from which he later deduced its origin and development.

Samuel Franklin Emmons went on to a distinguished career as a mining engineer, and those who are interested will find his eulogy in the *Transactions* of the American Institute of Mining Engineers (1911). I have called this ascent an anticlimax; yet it was that only in the sense of timing. The first scientific knowledge of our great volcanic peak resulted from the investigations of Emmons and Wilson — their ascent *was* important.

And so, Mount Rainier was at last certainly conquered, and doubly so; once in the name of whatever tempts men to the heights, and once in the name of Science.

CHAPTER III

SOLITUDES FOREVER BROKEN

1871-1884

After the two successful ascents in the year 1870, Mount Rainier remained undisturbed for twelve more years. Yet as the faint trail up the Nisqually River grew dimmer, events were maturing which would abruptly end the isolation of the mountain. And here again, as elsewhere on the frontier, it was the coming of the railroad which generated change, truly providing that panacea the trans-Cascade wagon road had failed to bring two decades earlier.

By 1873 THE NORTHERN PACIFIC RAILROAD had been built northward from the Columbia River to Tacoma, the western terminus created on Puget Sound. From there the plans called for construction of a "Cascade Division" which would cross the mountains to join the main line at the mouth of the Snake River, but financial embarrassment prevented an early beginning and it was not until 1875 that construction could be resumed. Then the necessity for an immediate return from the investment caused a temporary abandonment of the original plan in favor of a branch line to tap the newly discovered coal fields on the Puyallup River. The branch line was built and at its end the coal mining town of Wilkeson appeared in 1877.

Control of the Northern Pacific soon fell to Henry Villard who saw that the railroad would have to dominate the coal fields to be successful. He began by organizing a geological exploration to determine their extent, employing a capable and energetic young geologist to direct the field work.

Bailey Willis began with a reconnaissance of the forested country to the west of Mount Rainier, preparing in the

58

course of his work, a map which showed the names of its principal glaciers.[1] In 1881, just as the mountain was receiving its first true tourists,[2] he constructed a trail southward from Wilkeson to the north fork of the Puyallup River where he established "Palace Camp"— a great, barn-like building of logs twenty by forty feet. From his headquarters there, Bailey Willis prospected for coal fields from Carbon River to the Busywild, and secured control of worthwhile deposits by placing his men on "claims" which the railroad took over as soon as the titles were proven.

But Bailey Willis was more than a specialist in land grabbing. He recognized the tourist possibilities in the alpine parks on the north and west slopes of Mount Rainier, and he proposed that the Northern Pacific Railroad enter the tourist business by constructing a spur trail into what is now Spray Park. The revenue-hungry railroad approved of his plan (they would have little in it anyhow); so the trail was built. The wilderness barrier surrounding Mount Rainier was breached at last.

However, more came of that proposal than a mere route to the mountain. The railroad had begun to orient its advertising toward the promotion of tourist traffic, playing up the scenic attractions of the Pacific Northwest, and Mount Rainier was easily fitted into the scheme. An opportunity to identify the great peak with the railroad's western terminus was quickly seen in the name Tacoma; accordingly, the *Northwest Magazine* for March, 1883, made the following announcement:

> The Indian name Tacoma will hereafter be used in the guide books and other publications of the Northern Pacific Railroad Company and the Oregon Railway and Navigation Company instead of Rainier which the English Captain Vancouver gave to this magnificent peak when he explored the waters of Puget Sound in the last century.[3]

The Tacoma *Ledger* immediately adopted the new name and used it consistently thereafter, which led to a rapid acceptance by the people of Tacoma. The word which Theodore Winthrop supplied through his imperfect understanding of an Indian adjective, was about to become the focus of a mighty controversy.

In 1883 a man came north to climb a mountain. George B. Bayley certainly had the mountain fever, for he had already climbed Mounts Whitney, Shasta, Lyell, Dana, Hood, Lassen, and Pike's Peak, having been a companion of John Muir on some of his scrambles in the Sierra Nevada.[4] Bayley's enthusiasm for the unspoiled wilderness met with the entire approval of that great naturalist, who has left us a vivid impression of him joyously halooing each grand vista so that "wild echoes were driven rudely from cliff to cliff."[5] This wiry, little man from California has been called one of the most remarkable climbers of the time,[6] yet little else is known about his early years, except that he was once a sailor.[7] At the time of his appearance in this narrative he was listed in the Oakland city directory as a manufacturer of incubators.[8]

While on a business trip to Portland in July of 1883, Bayley learned of the trail recently constructed from Wilkeson to the northwest slope of Mount Rainier, from where it was said the summit could be reached in one day. Of course, he was off at once to make the ascent only to find the information false, for Bayley was too experienced a climber to make an attempt on the forbidding slopes that rise above Spray Park. Though disappointed, what he saw increased his desire to climb the mountain; so he went to Yelm and looked up Van Trump, who agreed to guide him to the summit by the route used by himself and General Stevens in 1870.

Together, they persuaded the old pioneer, James Longmire, to pilot them to the foot of the mountain, but he would not go until his harvest was in — a delay of two weeks. Bayley passed the time with a visit to Victoria on Vancouver Island, after which he returned to get together an outfit for the journey to Mount Rainier. Van Trump says they were well equipped and provisioned (listing flour, bacon, coffee, English breakfast tea, chocolate, canned Boston baked beans, codfish balls, beef, potted tongue, Leibig's extract of beef, jerked venison, oatmeal, potatoes, butter, sugar, condensed milk, canned apricots, peaches, and pears).[9] He proudly adds that no tobacco was taken and only one small flask of liquor for emergency use. At the last moment William C. Ewing, the son of Congressman Thomas Ewing of Ohio, joined the party, though he did not commit himself to climb to the summit.

When preparations were nearly completed, William Packwood and a group of woodsmen returned to Yelm after failing in an attempt to retrace his old trail. He reported the route impassable and so alarmed Mrs. Longmire that she attempted to dissuade her husband from going on the trip, clinging to him and saying, "Jim, you just shan't go!" But Jim did go.

On the tenth of August they left Yelm, each man riding a saddle horse, and with the camp gear and the provisions on two packhorses. The party crossed Nisqually River near the present bridge at McKenna, then followed a fair wagon road nearly to Mashel Prairie, which was then on the edge of the settlements. There they stayed the night at the farm of a Klickitat Indian known as Henry, though his real name was "Soo-too-lick."[10] He had settled on the prairie in 1876, developing a prosperous farm with fenced fields planted to wheat and vegetables, a log house where he lived with his three Indian wives, and a barn in which the weary traveler could always find a bed in the hay.[11]

Indian Henry had taken up the "Boston" ways of his white neighbors in earnest, and occasionally, to a comical extent. Van Trump tells of seeing posted on the door of the barn a notice which he thought concerned a school or road meeting, until he read the following: "Notice — Any one entering here is liable to instant death, as I have set spring guns inside. To all whom it may concern. Indian Henry." It probably was only a bluff, for he was really a kindly man.

Most of Indian Henry's fifty-five years had been spent in the forests about Mount Rainier, so Longmire was eager to get his services as a guide to the mountain. After a long talk in Chinook jargon (Henry spoke little English), the woods-wise old Indian agreed to act as guide for two dollars a day. He also said that he knew a route, thirty miles shorter than the one used in 1870, by which he could take the horses to the snow line. Though doubtful, the party agreed to try it.

The first difficulties of the trip were met soon after leav-

61

ing Mashel Prairie. The trail passed through an area still smoldering from a recent forest fire, and the yellow jackets were in a blindly revengeful mood. The poor horses were stung into snorting, kicking stampedes which were often stopped only by thickets or barricades of down timber. The riders also took their share of stings, to which the trouble of repacking the horses was frequently added. Van Trump says the experience brought out a virtue of the party; its members did not swear. Four hours of hard travel, relieved only by periods of vigorous axe-work, brought them to the clear waters of Mashel River near where the town of Eatonville now stands.

The route was better after leaving the river for they had entered the primeval forest. Four more sweaty hours of working the horses around and over the down timber that covered the forest floor beneath giant trees, brought them to a small stream running south to the Nisqually River, where they camped that evening on a grassy bar. There they were denied sleep or rest by swarms of gnats and mosquitoes. Mud failed to make their exposed skin any less vulnerable and blankets proved to be poor protection. Bayley says the insects got into the food without improving its flavor, and he describes his tea as a "nauseating puree of gnat." Camp life was rugged but they became inured to it, for each night was the same until the snow line was reached.

On the third day the tedious journey was continued through a forest which seemed to grow denser and more cluttered with fallen trees. Occasionally the lead horse would stir up a nest of yellow jackets; then Indian Henry would cry out the warning, "Sojers! Hyack claterwar!" and off they would go at a gallop.[12] A creek known to Henry as "Sakatash," and to Longmire as Silver Creek, was reached by 3 p.m.[13] They encamped at Copper Creek at six, after crossing a long stretch of old burn.[14]

On the morning of the fourth day, the trail followed the Nisqually River closely, fording it several times at places Bayley thought none too safe because of the boulders rolling along in the brawling torrent. It was soon necessary to turn Ewing's exhausted horse loose, after caching the saddle and bridle. Near noon they passed the point where the old

Packwood trail turned toward Bear Prairie, and from there on they had to blaze a new trail up the river. That night they encamped on the riverbank near what Bayley describes as "soda and iron springs of great variety."[15] The horses had good grazing even if their riders fared no better than before.

The following morning, just as they were mounting to ride, the pall of smoke that had lain over the country for two months lifted to give them a fine view of Mount Rainier, which lay northeast by compass from their camp. The course on this, the fifth day, was up the rocky bed of the river, with many crossings of the stream, until 11 a.m., when they forded to the east side within sight of the glacier. Midway in the crossing, some yellow jackets, nesting on an island, made a sortie on the horses, causing a near panic.

The party was now at the foot of an abrupt, timbered ridge, which all but the guide doubted could be ascended by the horses. Indian Henry proposed to cut a zigzag trail to the top, and it was begun while Bayley built a fire and started lunch. Having some spare time, the cook went off to examine the terminus of the glacier, which he found to be four hundred feet high, "contained between polished walls of grayish-white granite." He was gone too long, for another member of the party came down from the trail and finished the cookery, but not without moving the "kitchen" to avoid another yellow jacket nest.

After lunch the horses were resaddled and driven up the new trail, which took an hour and a half of toil. On top the ridge there were no more down logs to clamber over or avoid, the underbrush decreased, and they soon entered beautiful meadows set with clumps of alpine fir — the area now known as Paradise Park. The snow line was reached about 4 p.m. and a base camp was established on a flowery ridge at an elevation of 5,800 feet.

The plan was to push as high on the mountain as possible the next day, in order to set up a "high camp" from which the ascent to the summit could be made unimpeded by overnight equipment. Accordingly, on the morning of August 15, the packhorses were led over the snow for three miles, loaded with provisions, blankets, and a large bundle of dry sticks.[16] When the footing became too precarious for

63

the horses, they were unloaded and taken back to Indian Henry, who had refused to go all the way with them. The guide then took the horses back to the base camp, but not before trying to dissuade his friends from going higher. Being no more successful than Mrs. Longmire, he closed with, "Well, Longmire, closh nanitch, closh nanitch," departing in a gloomy, frightened mood.[17]

Shouldering packs of about twenty-five pounds apiece, the climbers toiled up the snowfield, occasionally passing around a crevasse, until a rocky cliff having loose stone at its base, was reached.[18] At that point, Bayley, Longmire and Ewing were so fatigued from the exertion and the heat that they wished to encamp, despite the arguments of Van Trump in favor of climbing higher. The bundle of firewood had been left behind, but two of the party went back for it while the others rested among the rocks.

A few sticks of the precious wood were used to cook the evening meal, then each climber wrapped his pair of blankets about him and lay down in a hole from which the rocks had been thrown. Though the view was magnificent (with shadowy ridges appearing above valleys shrouded in smoke, the whole overtopped by the great white cones of Mounts Adams, St. Helens, and Hood floating on the southern horizon) they did not stay up to admire it, for the temperature dropped from the 94° of midafternoon to 34° at sunset.

After a night made uncomfortable by the cold and the noises of the nearby glacier, the climbers awoke at 4 a.m. to find the top of the mountain obscured by a mist from which sleet was falling. They feared it would be impossible to continue the ascent, but by five the sun had burned away the fog, revealing a fine, clear day. Breakfast was quickly prepared, coats were rolled for carrying knapsack fashion, food for two meals was put in their pockets; then the extra items were distributed among them, and they took up their alpenstocks to begin the ascent to the summit.[19]

The route led off across 300 yards of hard snow to a dark ridge of loose rock[20] extending north up the mountain on a steep slope to a junction with a large mass of perpendicular

64

rock.[21] Two hundred feet of hand and foot climbing up the ridge convinced Ewing he did not care to climb Mount Rainier, so he handed over the barometer and returned to high camp. The others continued along the ridge, Van Trump noticing a considerable difference in it due to sloughing of the side next to the Nisqually Glacier during the interval since he had climbed that way with Stevens. They could no longer follow the crest the entire length, but were often forced to cut steps along the steep snow slope on the east side.

Reaching the great cliff now called Gibraltar Rock at 10 a.m., the climbers passed cautiously around its western face on the narrow ledge, holding to projections of the rock at some points. Rocks dislodged by their feet bounded noisily to the surface of the glacier far below, and once a mass of debris from the top of the cliff crashed onto the ledge ahead of them. At another point a huge, greenish icicle completely blocked the ledge, requiring hard work with the hatchet to cut a way by it.

Arriving at the chute formed by the junction of the glacier with the wall of the cliff, the climbers found it necessary to cut steps up the icy incline under the urging of volleys of rocks from above. The top of the chute was reached at noon, seven hours after leaving high camp.

Easier traveling had been expected on the slope leading on up to the summit; instead the surface of the snow was found to resemble frozen billows with crests three feet high, and progress among them was accompanied by many falls into the icy hollows. Van Trump says twenty-five feet of that exhausting travel was the limit of endurance, so that it was necessary for the rear man to move up to the lead every few minutes. A northwest course taken from the top of the chute was held except when passing around several large crevasses which lay on the route. At 2 p.m. a ridge of bare rocks could be seen above, and at 3 p.m. they stood upon the rim of the east crater, exposed to the strong north wind from which the bulk of the mountain had previously shielded them.[22]

Van Trump noticed that the level of the snow in the crater was considerably lower than it had been in 1870, with the inside of the crater rim bare and clearly defined. Struggling on against the wind, the party passed around the south and west sides of the large crater to the dome-like mass of ice which stands on the ridge between it and the smaller west crater. A hole was gouged in the ice on the highest point, the flagstaff, cut from a scrubby alpine fir, firmly set, and a flag was fastened to it to pound and snap in the violent gale blowing that afternoon of August 16, 1883.[23]

After spending an hour exploring the two craters, the climbers realized it was too late in the day to descend the mountain as planned. They then thought of visiting the unexplored North Peak before dark, but decided against it because of the wind. Probably it was at that time that the lead sheet marked, "August 16, 1883 G. B. Bailey P. B. Van Trump W. C. Ewing James Longmire," was placed near a steam vent on the edge of the west crater, though neither account of the climb mentions it.[24]

Faced with the necessity of spending a night on the summit, Van Trump searched for the cavern which had sheltered Stevens and himself in 1870. He found it quite different, for less than a third of the cave remained due to the decrease in the level of the snow. It was a shallow hole eighteen feet in diameter with the stones of the wall that had formerly enclosed the steam jet now scattered about in it. But the jet was still active, so Van Trump and Longmire proceeded to wall in a space around it sufficient for them to lie down on, while Bayley used his alpenstock to level off a path seven feet long outside the wall. Lacking an appetite for food, they settled themselves for the night.

Van Trump and Longmire, who lay down inside their low barrier with the steam vent between them, were soon damp from the hot vapor. Bayley had decided to walk to and fro on his short path, hoping in that way to keep both warm and dry, but the temperature fell to 20° and the cold, together with fatigue, forced him to lie down about midnight with his companions. There he blistered his fingers pulling hot rocks out of the vent to warm his feet with. In spite of his misery, Bayley was able to appreciate the sparkling beauty created by the bright moonlight in

66

that windy pit. By morning all had suffered enough; the nauseating odor of the steam, the alternate cooking and freezing, and their cramped positions made them willing to brave the unrelenting wind.

Fearing a storm might be developing, preparation for the descent was begun about seven o'clock. A few more caulks were hammered into their shoes, then they exercised violently within the crater. The temperature was 25° when they left their shelter at 7:30 a.m. Before they had gone one hundred yards their clothes were frozen stiff, forcing them to take cover behind some large rocks. Crouching there, they discussed the best way to get to the east side of the large crater where there would be shelter from the north wind, which Bayley estimated at 100 miles per hour. One was for crossing directly over the crater, but the others thought the partial protection which would be gained by following around the inside of the north rim was essential.

Bayley started in that direction, making only fifty yards before he fell from cold and exhaustion. Longmire fell several times and Van Trump lost his frozen hat to the wind. Crawling and staggering from one sheltering rock to another, they reached the point of descent and protection from the wind at 9 a.m.

The sun and the exercise soon warmed them as they traveled down the slope toward the top of the chute, where they arrived at 10 a.m. The rope was used to let them down its icy slope, the top man assisting the other two down the full length, after which he would double it over ice knobs and move down by half lengths. No sooner were they safely past the ledge than it was bombarded with rock. Passing on down the ridge, they arrived at high camp at 1:30 p.m., to find Ewing anxiously awaiting them.

During his lonely vigil at high camp, Ewing had generously denied himself fire and had eaten very little in order that there might be warmth and food for the returning climbers. He soon had tea, bread, and meat ready, and it was devoured greedily by the famished trio, who named the place "Camp Ewing" in recognition of the thoughtfulness of their friend.

After a short rest in the sun, the descent to the snow line

was begun at 3 p.m. Shortly after starting a thick fog enveloped them, forming out of a clear sky; however, they were able to follow the tracks made while ascending, reaching the base camp at 5 p.m. It was obviously abandoned. The tent had been carefully ditched, with the equipment stacked inside and covered, where it was protected by wooden scarecrows placed at the ends.

Some revolver shots were fired, a great deal of whooping was done, and later that evening, Indian Henry returned. He had been so certain the climbers were going to their deaths he had taken measures to protect their property, before moving the horses and his "ictas"[25] to a more favorable campsite (from which he had intended to set forth the next day for the settlements to tell of the tragedy he assumed had occurred). Henry was genuinely glad to see them — he had a "closh tumtum"[26] for he was thankful to be relieved of his onerous responsibility.

The next morning there was a flurry of snow and the little stream near camp had ice on it, which induced the climbers to lie abed later than usual; even so, the horses were packed for the return trip by nine o'clock. They made better time, four days instead of five, but otherwise it was much the same—yellow jackets by day, gnats and mosquitoes at night, with Longmire and Van Trump half blind from an eye irritation they attributed to the steaming in the summit crater.

The party reached Yelm safely on August 21, 1883, being "much bruised, blistered, and sore as to their persons, and much torn and dilapidated as to apparel," as Van Trump put it. Though a little disappointed at not reaching the North Peak, they were satisfied with the trip. Van Trump felt it vindicated him in the eyes of some neighbors who did not believe he reached the top in 1870; Bayley felt he had at last climbed a mountain worthy of the effort; and Longmire found he was still a good man despite his sixty-three years of hardships. Mr. Ewing, the good companion, went his way and will not reappear in this narrative, but the others will be met again at Mount Rainier.

George Bayley, the man who came north to climb a mountain, had accomplished a great deal more than that.

He had been a companion of John Muir, sharing the beliefs and feelings of that broad-visioned man to such a degree that he was bound to propagate them. Van Trump and Longmire returned with the germ of an idea which probably came from the campfire talk of Bayley, for each was convinced that Mount Rainier was the people's heritage and each worked in his own way to make it so.

The first evidence of the new idea appears in the article Van Trump wrote for the Olympia *Transcript* soon after his return. In it he says:

> Rainier with its varied and beautiful surroundings, is destined to become a favorite resort for tourists, pleasure seekers and invalids, and the people of Olympia and Thurston county have it in their power to make the route via Olympia and Yelm, and Nisqually river, the popular one to the mountain.

From that start, Van Trump wrote ceaselessly and well to popularize Mount Rainier.

James Longmire's vision took a different form. In it there was a trail to the mountain, with a hotel for the accommodation of the visitors he expected. It would come to be, but not quite yet.

The Northern Pacific Railroad was completed on September 8, 1883, near Gold Creek, Montana. The spike-driving ceremony gave the astute President Villard an excuse to show off his railroad, so a special train brought a number of important people to the Pacific Northwest on an expense-free tour. Always mindful of his German and British stockholders, they were well represented on the excursion. Among them was a scientist, whose interest in the glaciers of Mount Rainier led to organization of a side trip to the mountain.

Thus, it fell to Bailey Willis to guide a small party, including Senator George F. Edmunds of Vermont, over his trail to the beautiful alpine meadows now known as Spray Park, where they ascended the snowbanks to what was called "Observatory Point." Two members of the group, Professor Karl Zittel, a German geologist, and James Bryce, the author of *The American Commonwealth* and later the British ambassador to the United States, were so impressed by what they saw that they wrote Villard a joint letter

69

recommending that Mount Rainier be set aside as a national park.[27]

Since a portion of that letter is yet available to us, it is worth while to consider what they said:

> The scenery of Mount Rainier is of rare and varied beauty. The peak itself is as noble a mountain as we have ever seen in its lines and structure. The glaciers which descend from its snow fields present all the characteristic features of those of the Alps . . . are in their crevasses and seracs equally striking . . . We have seen nothing more beautiful in Switzerland or Tyrol, in Norway or in the Pyrenees, than the Carbon River glaciers and the great Puyallup glaciers . . . The combination of ice scenery with woodland scenery of the grandest type is to be found nowhere in the Old World, unless it be in the Himalayas, and, so far as we know, nowhere else on the American Continent.
>
> We may perhaps be permitted to express a hope that the suggestion will at no distant date be made to Congress that Mount Rainier should, like Yosemite Valley and the geyser region of the Upper Yellowstone, be reserved by the Federal Government and treated as a national park.

The pearl of great price had been found, yet ten years would pass before that discerning testimony was heard.[28]

Henry Villard completed the Northern Pacific mainline, but the effort brought the railroad to the verge of bankruptcy. As a consequence, that giant of American finance was forced to resign the railroad presidency January 4, 1884. Of course, the new management had new ideas, and Bailey Willis found his work for the Northern Transcontinental Survey brought to an abrupt end by a blunt telegram which informed him, "No bills paid after April 30 — Buckley General Manager." That gave him "Two days to close out the work of three and a half years! Two days in which to pay off more than a hundred men scattered in camps over a wide mountain area. Two days in which to save something from the general abandonment."[29]

Bailey Willis went on to greater things, taking with him an important knowledge of Mount Rainier and leaving behind a trail which would soon draw another party of climbers — a new type.

In 1884, three young men came south to climb a mountain. They, too, had the mountain fever, but unlike

George Bayley, they didn't know a thing about the sport; so they climbed it where he had refused to try. J. Warner Fobes, George James, and Richard O. Wells were as full of enthusiastic energy as the little boomtown of Snohomish from which they came. Fobes was the pastor of the Union Presbyterian Church at Snohomish, a ministry he undertook upon his arrival from Syracuse, New York, in September of 1883.[30] Whitfield describes him as "a young man fond of athletics, a believer in temperance," adding that he was active in community affairs and well-liked by his congregation.[31] His efforts to improve the educational facilities of the town are commemorated in the name of the Fobes School between Snohomish and Everett. It is likely that the Reverend Fobes organized the expedition to Mount Rainier as there are newspaper articles which show that he undertook several other trips into the Cascades before he returned to the East in 1887.[32]

Less is known about George James. His name appears on the Snohomish County census for 1885, where he is listed as twenty-five years old, single, white, a surveyor, and from Ohio. Although he was not listed in either the 1883 or the 1887 census, it is probable he was living in the vicinity of Everett as late as 1920.[33] As for Richard O. Wells, we must take the word of Fobes that he was a lawyer, for the name does not appear in any of the available records.[34]

On August 11, 1884, those adventurous young men left Snohomish for Tacoma on one of the little passenger steamers that swept gaily up and down the Sound — the Greyhound busses of that day.[35] They were off to climb Mount Rainier and quite excited over the prospect, for Fobes says they shared the opinion of most Puget Sound residents "that the mountain never had really been ascended." From Tacoma they traveled by train the following morning to Wilkeson, arriving there at 8:45 a.m. after a pleasant ride of two-and-one-half hours through oat fields, hop yards and raw clearings.

The climbers did not stop long in that small coal mining town.[36] After buying some flour and bacon at the one store, and eating a hearty meal of boiled beef and beans served them by the Irish women who ran the miners' boarding

71

house, they left afoot on the Bailey Willis trail under sixty-pound packs.

The trail passed southward from town, first rising over several broad, heavily timbered benches, then descending to Carbon River, which had to be forded as the bridge was gone. The river was passed at five o'clock and a night camp was made at the top of the ridge on the south side.

The next morning it was raining, so camp was not broken until 9 a.m.; even then it was no fit day for traveling and they camped again after four miles. But August 14th was much better. With an early start, Bark Town was reached at 11:30 a.m.,[37] and the little meadows now known as the Mountain Meadows were passed at three o'clock. From there on snowbanks were frequent along the zigzag trail to Crater Lake, where they camped for the night in the grassy meadows near the outlet.[38] At that pleasant camp they had a refreshing swim and a good night's sleep under four blankets.

Arising early the next morning the three climbers ascended the peak east of the lake before breakfast.[39] While they were resting on top, the clouds parted to give them their first view of Mount Rainier since leaving Tacoma. They had a good look at it and the surrounding country, then went back to breakfast and were soon on the trail again.

At the point now known as the Eagle Cliffs a short stop was made to enjoy the grand view from "a great rock a few steps from one side of the trail." It was their first close view of a glacier—a vista that Fobes called one of the most beautiful in America. Then on again, up a switchback course by the side of a waterfall, described by James as four hundred feet high,[40] to the trail's end in the alpine meadow a mile beyond.

Near timber line a base camp was established by constructing a rude hut with sod walls and a tent roof. That evening, as they sat by their night fire, all three were confident they would reach the summit of Mount Rainier next day.

Saturday morning, August 16, an early start was made up a ridge which appeared to lead directly to the top. The three climbers had selected the dangerous Ptarmigan Ridge for their first try. One of them carried a small axe and an aneroid barometer, another the lunch, and the third had one hundred feet of light rope; while all had six-foot, steel-pointed alpenstocks of ash and wore boots with inch-long caulks.

The grassy slope above camp soon gave way to rocky ground, from which the climbers passed to permanent snowfields, where the traveling was easier. On that firm surface, made solid by weeks of thawing and freezing, their progress was rapid during the cool early hours, but with the heat of day came fatigue which slowed them to a weary plodding. That toilsome gait was further slowed by the appearance of crevasses. So much time was lost in passing them, the barometer indicated only 10,000 feet at noon.

The climbers continued up a rocky spur to 11,000 feet, to be brought to an abrupt halt on the "rim of a great crater basin," cut off from the main mountain by rocky walls and impossibly steep slopes of snow and ice seamed with crevasses.[41] Wells pushed on a little farther in a hazardous search for a route up that nearly sheer two thousand feet between them and success, but found nothing they could negotiate. After shivering for a time on the sunny side of some rocks, the climb was called off. Fobes says that as they sat around the campfire that evening, they admired the moonlit beauty of the mountain — but it was admiration "accompanied by a respect for its ruggedness we had not the night before possessed."

Sunday morning they all slept late, with Fobes the first one up. He was busily chopping firewood when he noticed what appeared to be a "huge" collie dog trotting over the snowbanks toward the encampment. At closer range the animal proved to be a large wolf, and Fobes called to his companions to hand the rifle out of the tent. As they brought the gun out to him the wolf ran off to about eighty yards, where a hastily-aimed shot struck him with no more immediate effect than to send him off across the valley leaving a bloody trace.

After a late breakfast, the three went exploring, each according to his fancy. Fobes went northward, following down a beautiful, flower-filled valley which brought him to Carbon Glacier where it emerged from the rocky gorge which has since been nearly evacuated.[42] A hundred-foot scramble put him on the debris-covered top, which he likened to "a dried worm with its skin all cracked open, only on a somewhat larger scale"; thereafter, some rough traveling, replete with good experience in route finding on a glacier, brought him to a little hill on the east side (possibly the one now known as Goat Island Rock). From that vantage point, Fobes examined the north and northeast slopes of Mount Rainier, finding what appeared to be a possible route to the summit. He also found a better route across the glacier and used it to return to camp.

James was already there, having concluded his ramble with a ludicrous encounter with a black bear. Of course, the hero is silent about it, but Fobes tells the story with obvious relish. He says:

> Coming home over a high ridge, he saw an immense bear down five hundred feet in a valley, and, as he had the rifle with him, he concluded to give Bruin a shot. He started down, but after descending about half way came to the conclusion that the bear ought not to be so rudely disturbed, and struck out for camp. We never could determine whether the fact that it was Sunday, the depth of the valley, or the size of the bear was the most instrumental in bringing him to this conclusion.

No doubt the faint-hearted hunter paid dearly for not keeping that episode a secret.

The last one back to camp was Wells, who came in pale and tired, as the others were busy with their supper. He appeared so glad to see his companions again that they knew something unusual had happened. Finally, Wells told them he had gone onto the Carbon Glacier in the great cirque near its head, where he had fallen into a crevasse from which he never expected to escape alive. He managed to get out by climbing up a narrow place while hanging masses of snow fell near him, threatening a sudden, chilly burial.

The information gained by Fobes on his reconnaissance across the Carbon Glacier was put to advantage on the following morning, when they broke camp early to move eastward around the base of the mountain. Before they reached Carbon Glacier, a large mountain goat was surprised where he was feeding between the snowbanks. Fobes, who was ahead with the rifle, finally brought down the awkward-gaited animal on the third shot, but neither the accomplishment nor the trophy was as satisfying as the hunter had expected. Fobes dismisses goat hunting by offering his opinion that "cow-shooting in a big pasture might be as difficult." The disappointment was later heightened when their cookery failed to rid the meat of its "billy" flavor and seventeen-year toughness.

Laden with such meat as they could carry along with their already heavy packs, the party crossed Carbon Glacier, then passed through the beautiful meadows now known as Moraine Park to a camp site on the ridge between Carbon and Winthrop glaciers, at an elevation of 6,100 feet by the barometer.[43]

From the new encampment, the spectacular avalanches which course the nearly vertical face of the Willis Wall were in full view. After sunset, frost action began dislodging large blocks of ice from the great cornice above, sending them crashing and grinding down the 3,000-foot cliff to break into a cloud of snow-dust which settled like a pall of smoke on the ice in the cirque below. The delayed reverberations of some of the mightier avalanches were occasionally able to awaken the soundly sleeping campers during the night.

On August 19th, after "disposing of the luxuries of life," as James so quaintly describes their simple breakfast, a start was made up the mountain by the rocky spur which forms the eastern edge of the Willis Wall. All went well up to an elevation of 9,000 feet, where a perpendicular face prevented farther progress up the ridge. The only alternative was to drop down to the surface of Winthrop Glacier and make their way up it to the summit. The best place for that was nearly a mile behind them, but rather than lose elevation, they attempted to descend to the glacier by a connecting tongue of ice lying on a slope of fifty degrees.

It was a bit of foolhardiness which nearly cost Fobes his life. The job of cutting steps with the axe across several

75

hundred yards of the slope immediately above a wide crevasse seemed much too slow for him, so he pushed ahead, depending entirely upon shoe caulks for footing. A few steps were safely made, then Fobes slipped, fell on his back and began a rapid slide toward the crevasse. Retaining his presence of mind, he grasped his alpenstock with both hands above the point, then rolled over and dug the tip into the icy surface to drag himself to a stop just short of the chasm. James says, "After commending the boy for the coolness displayed, when imminent death was staring him in the face, we passed on."

When at last upon the glacier, the traveling was easier, though many crevasses were encountered on that route. Two of those were extremely difficult to pass, for the downward slippage of the ice had left the upper wall standing twenty-five feet above the lower, and strenuous axe-work was required. The climbers reached an elevation of 12,000 feet by 11:30 a.m. Not far above that level the air became very cold, and the barometer ceased to operate, though designed for use up to 16,000 feet.[44] A lunch stop was made at one o'clock in the shelter of some ice pinnacles, but it was so uncomfortable there they soon plodded on. Their exertions in that cold, rare atmosphere began to sap their energy and rest stops became necessary at more frequent intervals.

By 1:30 p.m. James was exhausted; he was pale, faint, and his feet were freezing from poor circulation. Half an hour was spent rubbing his feet to restore warmth to them; then Wells, who was reluctant to give up the climb so near the top, pushed on toward the summit followed by the tired and unenthusiastic Fobes. After going a hundred yards they looked back and saw James staggering around, gasping for breath, where they had left him. Abandoning the climb, they all descended and James recovered rapidly after passing down a thousand feet. A better route was found on the return to camp, sparing them many difficulties.

The second failure to reach the top left the climbers quite depressed, for their provisions were down to a two days' supply of flour and bacon. As Fobes knelt by the little stream mixing "slapjacks" that evening, Wells came down and said he had made up his mind to reach the summit the next day, even if he had to go alone. Fobes agreed to accompany him on another try, and the next morning, James also decided to go up again.

Leaving camp at 7 a.m., they traveled more rapidly, reaching their lunch stop of the previous day by noon — a precious hour gained! Though the weather was hot at the start of the climb, it again became very cold above 12,000 feet, yet they did not suffer from it as before. Not far above their former "highest," a small point was rounded, putting the summit in full view, and Fobes says, "The inspiration of success was upon us and overcame our fatigue, though we had to stop every five minutes to catch a full breath."

The route followed brought the three climbers into the saddle on the ridge connecting the North Peak with the true summit. By 2:30 they were within 300 feet of the top, where they linked arms for the remaining climb onto the snow dome now known as Columbia Crest, arriving equally "first."

Fobes says they were unaware the mountain had ever been climbed before, and had just given three wild cheers to celebrate the conquest when they saw a "most common, scrubby-looking" stick protruding from the snow of the dome. It was the stick placed there by Van Trump on the climb with Bayley and Longmire the previous year,[45] and it served as a warning "not to claim too much for themselves." According to James, they put a message in a tin box, fastened it to the stick, and then began an exploration of the summit.

Near a steam vent where the climbers huddled to warm their hands, a sheet of lead four inches by eight inches, bearing the names of the Van Trump-Bayley-Longmire party, was found. James says they put their names on the back and replaced the record.[46] While on the summit, they were surprised to find a large butterfly fluttering over the snow. They also noted a fact which is always a great disappointment to the climber on his first ascent. Though the weather was bright and clear, without a trace of cloud or fog, the view was not as good as it had been at lower elevations, which prompted Fobes to write, "in this respect

. . . the sight is much less impressive than one naturally expects."

During the hour they spent on top, the features of that vast panorama were imprinted on their memories: the mountains fading into a blue dimness overtopped by distant sentinel peaks; the Sound, with just the trailing smoke of a steamer to mar its glistening sweep of waters northward to the Straits of Juan de Fuca; and that mighty carpet of forest, only occasionally broken by prairies and man-made clearings, and threaded by the rivers originating in the glaciers radiating from the summit whereon they stood. Taking that with them, the downward journey was begun.

Though the ascent required seven and one-half hours, they managed to descend in two. The abandon with which they slid down those steep, dangerous slopes, was the crowning recklessness of their trip. James says they did it this way:

> Taking our Alpine staffs firmly in our grasp and letting the sharp ends drag behind us in the snow and ice, serving as brakes, we slid rapidly down the mountain . . . In some places we would have a smooth surface of several hundred feet to slide, at angles from 45 to 60 degrees, and again having to stop short to cross some deep crevice or to descend some precipice.

And so, with their guardian angels hovering over them, they reached the encampment safely at six o'clock. The ascent of August 20, 1884, the first indisputably successful climb of Mount Rainier by a route other than Gibraltar, was complete.

They were "satisfied, jolly, and hungry," to which was soon added a sensation probably new to all of them — snow blindness! That night their eyes were so inflamed they could not sleep and were forced to apply handkerchiefs wet in cold snow water to get relief (temporary relief at that, for the pain returned as the cloths warmed). Five days of continuous exposure to the reflection of the summer sunlight on snow rendered them nearly blind.[47] They were hardly in condition to travel the next morning, but the depletion of their provisions forced them to start.

Despite their condition, the return trip was uneventful. The successful mountain climbers reached Snohomish on

August 25, sunburned, peeling, and "satisfied with mountain climbing for the season."

As the enthusiastic young men from Snohomish pass from view, let us take a look at the other side of the mountain, where we can watch James Longmire turning dreams into reality. Soon after the ascent of 1883, he approached Bayley and Van Trump with a proposal to build a good trail up the Nisqually River and develop the mineral springs they had found. But Bayley wasn't interested, and Van Trump claimed he had no money to spare for such a project.[48] So the old pioneer decided to go it alone.

With some of Indian Henry's relations to help him, he cleared the rough trail up the Nisqually River from Mashel Prairie to the mineral springs. By the end of fall, 1884, he had built a rough cabin near the "soda spring." Family tradition says that he was accompanied by his wife, Virinda, on that trip, but it seems improbable.

And so, the wilderness barrier surrounding Mount Rainier was breached, and there was a lodgement deep within. Two passable trails and a cabin — harbingers of things to come. Never again would there be silence in that forest for any length of time.

CHAPTER IV

LOWER GARDENS OF EDEN

1885-1888

The Northern Pacific Railroad made Mount Rainier accessible, but that was not the whole of its effect. Railroad and steamship company advertising in the East and Europe, during the decade prior to Statehood, resulted in population growth unequalled anywhere else in the United States. Many of the new settlers were not satisfied with what they found in the towns, on the prairies, and along the rivers, where lay the loose web of settlement about Puget Sound. They pushed out into remote valleys and penetrated the foothills of the Cascades in search of homesteads — forcing the wilderness to retreat before winding dirt roads and the little towns that followed them.

THE TRAIL WHICH JAMES LONGMIRE OPENED to his mineral claim at the Soda Springs soon became known as the "Yelm trail." It had the immediate effect of lodging an outpost of civilization in the valley of the Nisqually River. During the summer of 1885 a man by the name of James B. Kernahan located a homestead in that part of the valley known to the Indians as *So-ho-tash,* the place of the wild raspberry, a name which local usage would later corrupt to Succotash Valley. There Kernahan settled, building a comfortable log house for which all furnishings were brought in by packhorse. Unlike James Longmire, who only occupied his claim during the summer and fall, Kernahan moved his family to the "Palisade Ranch," as he called it, and they lived there the year around, twenty miles from the nearest neighbors.

We also know for a certainty that James Longmire brought his wife, Virinda, to the rude cabin at the "Springs" that summer, where they had at least one guest — a Mrs. Jameson of Olympia — who probably was the first paying guest at Mount Rainier.[1] While the wildflowers were at their best, the Longmires followed Indian Henry's switch-back trail to the alpine meadows above Nisqually Glacier. David Longmire says, as they rode into the park, his mother exclaimed, "O, what a paradise!" and Mrs. Jameson answered, "Yes, a real paradise."[2] Time would turn that spontaneous remark into a place name presentive of all the alpine regions of Mount Rainier.

A place name of nearly as great importance secured a lodgement the following summer when a party consisting of Charles E. Kehoe, Charles A. Billings, and George N. Talcott, of Olympia, camped at Paradise Park under adverse conditions. On August 12, 1886, the clouds parted giving them a superbly-framed view of the mountain, whereupon they called their campsite "Camp of the Clouds."[3] The name was poetic and appropriate, holding the public fancy until the era of the hotel and lodge robbed it of significance.

There were other "tourists" afield during the summer of 1886. A group from Seattle came to Mount Rainier by way of the old Bailey Willis trail. Their unsuccessful attempt to ascend the mountain from the northeast is notable only because it included a Seattle school teacher by the name of Edward Sturgis Ingraham. The anti-Chinese riots in Seattle would soon give him an opportunity to use his organizing ability in the formation of a "home-guard," leading to a commission in the state militia and the title of "Major" Ingraham. Mount Rainier would see much more of him.

In the fall of that same year, a party of Yakima Indians gathered for a hunt in the Cascade Mountains to the west of their reservation. The people of that tribe had always considered the country over to and including the eastern slope of Mount Rainier as peculiarly their own, and the expedition was no different from those of former years, except that they invited a young white boy by the name of Allison L. Brown to accompany them.[4]

The party of about thirty Indians, with their white guest, crossed the Cascades by way of Packwood Pass, then moved up the Ohanapecosh Valley and onto the Cowlitz Divide. Finding no game there, in what should have been the best hunting ground, a move was made closer to the snow line, but with no better result.

In order to pass the time, several of the more adventurous suggested an ascent of Mount Rainier. That was agreeable with some, and seven or eight Indians started for the summit, accompanied by Brown. He says:

> From the contour maps I have seen since, my impression is that we continued to the end of the Indian trail on the Cowlitz Divide, and from there made for the lower end of Whitman Crest, skirting the end of the Ohanapecosh Glacier and from there to the ice field now called Whitman Glacier, crossing it and the ridge between the Whitman and Ingraham, and dropping down upon the Ingraham Glacier. At about 8,500 feet we crossed to the south side. We had used our horses as far as the near side of the Ingraham—much farther probably, than would be considered possible by a white man, and from this point we sent them back to our camp at timber-line.
>
> Continuing on foot, we followed up the west slope of what is now known as Cathedral Rocks on the Ingraham Glacier, making use of the well-defined goat trails. As I remember, there was a short distance of 40 to 60 feet where we were compelled to work ourselves along a ledge by gripping the side wall with our fingers, the ledge being very narrow, apparently just wide enough for the wild goat to travel over. After crossing this small strip, we found ourselves again on the glacier snow, and from there had an unobstructed, though rather steep climb over the snow to the top. We did not try to reach the highest pinnacle. The snow, as I remember, at that time was rough and granular, and the walking was comparatively easy. Most of the party wore the usual Indian moccasins and some of us had alpenstocks which we cut from the mountain ash and other shrubbery along the wooded spots. We took rations and axes, and carried one or two lariats to use in case of emergency, but never found it necessary to use them.
>
> In descending we tried to retrace as near as possible our own footsteps. Late in the afternoon we put up for the night at the base of the rock I have always believed to be Gibraltar. We found a rather sheltered place,[5] and the following morning descended to join the rest of our party and continue the hunting expedition. We were in the mountains approximately six weeks in all.

View of Rainier from Tolmie Peak, with Eunice Lake in the foreground. (Courtesy Mount Rainier National Park)

Snout of Carbon Glacier from the north side, by Bailey Willis, c. 1896.
(Courtesy Mount Rainier National Park)

Upper left: William Fraser Tolmie, Hudson's Bay Company physician,
fur trader and early botanizer at Mount Rainier.
Upper right: Philemon Beecher Van Trump at "Wigwam Camp,"
c. 1908. (Susan Hall Collection)
Lower left: Augustus V. Kautz, guided by Wapoety, almost succeeded
in scaling Mount Rainier in 1857.
Lower right: Hazard Stevens, youngest brigadier in the Union Army,
Congressional Medal winner and "Master of Rainier."

Upper: James Longmire, pioneer settler, Cascade roadbuilder
and trailblazer. (Courtesy Oregon Historical Society)
Lower: First photograph of Longmire Springs, by A. C. Warner, 1888
(probably August 11). (Courtesy Mount Rainier National Park)

Upper: 1905 visitors took National Park Transportation Co. cars to examine the grey snout of Nisqually Glacier. (Author's collection) *Lower:* The same view in 1955 shows marked shrinking of the glacier.

Upper left: Dashing Fay Fuller, teacher, persuasive Tacoma
editor and climber par excellence.
Upper right: Helen Holmes, one of the fourteen who climbed
to the summit on July 18–19, 1894.
Lower: (left to right) W. O. Amsden, Leonard Longmire, Fay Fuller and
leader E. C. Smith with 3-foot mercury barometer. (Author's collection).

Upper: Looking over Cowlitz Glacier from the Cowlitz Rocks.
(Curtis and Miller photo, #21809,
courtesy Washington State Historical Society)
Lower: Summit of Mount Rainier showing both craters,
from an aerial photograph taken by Lieutenant Hodgson in 1950.
(Courtesy Mount Rainier National Park)

Aerial view of Mount Rainier, with Mount Adams in the background. (Courtesy Mount Rainier National Park)

Brown adds that his companions were all agency-educated Indians and probably devoid of tribal superstitions. It is evident from his account that the ascent was incomplete, and how high these climbers actually went will probably remain unknown. Yet it is a curious coincidence that the Ingraham Glacier route was first used by the climbers in the same year that E. S. Ingraham looked from St. Elmo's Pass (which he named then) toward the glacier that one day would be named for him.[6]

It was also during 1886 that the German geologist, Dr. Heinrich Berghaus, published his *Physikalisch Atlas*. Part 6 (Gletscherkarte), included a map of Mount Rainier's glaciers as shown on the Bailey Willis map of 1880.[7] However, Berghaus took the liberty of changing the name of the Mowich Glacier to honor Willis, a substitution which was a continual embarrassment to the modest geologist until it was rectified by the United States Board of Geographic Names in later years. Even then, he was not left unhonored; his name was transferred to that tremendous cliff which backs the cirque of Carbon Glacier. It was a fortunate change, for the Willis Wall is nearer his stature and a more appropriate memorial to him.

Ingraham came back for another try at the northeast side of Mount Rainier in 1887.[8] With "several ambitious Seattle climbers" he reached the 13,800 foot elevation before being forced to give up the ascent because of a developing snowstorm "and what appeared to be an impassable crevasse in front." But he was gaining valuable experience, as well as enriching the map with picturesque names such as "Interglacier," "Battleship Prow" (now merely Steamboat Prow), and "Elysian Fields," which he originally intended for what is now known as Moraine Park. Yes, Ingraham had a knack for right names.

Glancing backward momentarily, the year 1882 saw the arrival of the family of Edward N. Fuller in the rapidly growing town of Tacoma — population 6,000 persons. They were probably pleased to find there was a public high school where twelve-year-old Evelyn Fay could continue her studies. When the high school closed after her second year, the young lady showed her resourcefulness by turning to school teaching to further her education.[9]

Her first school was at Yelm, where the new teacher probably boarded with the families of her pupils, according to the custom of that day. The resulting acquaintance with the Van Trumps seems to have developed into a special friendship, and Fay was greatly impressed by that gentleman's stories of his two ascents of Mount Rainier. He even came to her school on two occasions to talk to the children about the mountain.[10] The young teacher's interest became so great she made the trip to Paradise Valley in 1887, climbing to an elevation of 8,700 feet on the snowfields.[11] That was high enough to see Fred Plummer's flag snapping in the wind from the wire staff he had jammed into a crevice at his camp on Anvil Rock, his base for mapping the southern slopes. Unfortunately, there are no details of that trip, which may have been made as a guest of the Longmires. The important result of her visit was a heightened interest and a determination to some day do what no woman had yet done, and no man thought possible — to climb to the summit of the great peak.

With fall, there was school again to crowd out the memories of alpine meadows and rock-ribbed, glacier-clad heights. But not far to the south, a group of men were translating their memories of high places into something entirely new in the Pacific Northwest — an organization of mountain climbers. And so, on October 7, 1887, the Oregon Alpine Club came into existence at Portland.[12] It was an imperfect organization, with poorly defined objectives which would lead to wrangling and early dissolution, but for the time its influence was great and its accomplishments not inconsiderable.

That same year of 1887, John Muir received a proposal from the J. Dewing Publishing Company, of San Francisco, to edit and contribute to an illustrated work to be called *Picturesque California*. Muir was then forty-nine years old, married and well settled upon his California ranch; but his heart was not there, and his inspired writing had ceased.

Muir accepted the proposal and plunged with increasing enthusiasm into the new task, which brought him to Mount

Rainier and accomplished much for both of them. His biographer says of his return to the wilderness world he loved so much: "John Muir, in the fragmentary journals of these trips, resumed his writing. And he climbed Mount Rainier with something like the old ecstasy. 'I didn't mean to climb it,' he wrote to his wife, 'but got excited and soon was on top'."[13] Here was a man who had known the mountain fever before!

This should properly begin where John Muir boarded the "Pullman car" enroute for the Pacific Northwest. He picked up his friend, the artist William Keith and wife, at Lake Tahoe, and together they continued slowly northward collecting material as they went.[14]

On July 20, 1888, they arrived at Seattle, where they were the guests of Attorney General J. B. Metcalfe, at whose home Mrs. Keith was to stay during the Mount Rainier expedition. Several short excursions were made while the preparations for a mountain trip were completed; then some of those "ambitious Seattle climbers" we have already met were found to accompany Muir.

One member of the Seattle group was Arthur Churchill Warner, a young photographer. Shortly after his return from the expedition he wrote to his father, describing the trip in a folksy, carelessly composed letter that has fortunately been preserved. He had this to say about the start:

I was already to go to Alaska when a man came to me and said; "We want you to go to Mt. Ranier with us as special artist." At first I said I could not go as I had to go to Alaska but when they said I must go and that I would be well paid I said "All right" and on the morning of Aug. 8 at 3 o'clock my alarm went off. I turned over, rubbed my eyes and got up, I had everything ready and was soon on the way to the train.

Mr. Ingraham went with me instead of going to Alaska. We had taken our traps down the night before and now we only had a few small things to take. At the train we met the rest of the party, Mr. John Muir, a well-known writer and explorer on the Coast. It was he who wanted me . . . William Keith an artist of Cal., D. W. Gass, Chas. Piper, N. O. Booth, N. Loomis, one of the compilers of the Loomis text books, used in the colleges east, E. S. Ingraham and Yours Truly. We were all on time, save Booth who was late and got left. The train went to Tacoma, then 20 mi to Yelm, where we got off and found we

could not get our pack train till the next day, so we put up our
tent, spent the day in making negatives, sketches, etc. Mr. Bass
and myself slept in a barn that night on new hay, the others in
the tent and around the old rail fence. There are only two
houses there and so we did not get beds.[15]

John Muir had an introduction to P. B. Van Trump,
from their mutual friend, George B. Bayley. He wanted
the services of that veteran of two memorable ascents (who
was then the storekeeper and postmaster at Yelm), as guide
for the party. Muir calls him a "volunteer," but the matter
was not arranged that easily. Van Trump says, "though
my business and my wife being without help really made
it a dereliction of duty for me to leave home, they soon
talked me into the 'mountain fever'."[16] Arrangements were
also made with James Longmire to furnish pack and saddle
animals, and two men to handle them.

They were at breakfast on the morning of the ninth
when a freight train rattled by and Booth jumped off, completing the party. The pack train came up at 9:30, so the
work of packing that "cumbersome abundance of camp-
stools and blankets" on the seven tough little cayuse ponies
could begin.[17] The first animal loaded took off stiff-legged,
quickly scattering the pack and making more work for the
packers.

Of the two men furnished by Longmire, John Hays
passed without comment, and remains nearly unknown;[18]
but the other, Joe Stampfler, was an obvious youngster.
Muir was concerned over the animals and the "mere boy"
who was to handle them, but he was told the ponies could
stand up to their loads, and that Joe, though only fifteen
years old, knew the diamond hitch and had managed
pack strings before.[19] Joe's home was in Olympia, but he
did not get along with his father, preferring to live with
James Longmire's family, where he took care of the horses
and did chores for his keep. He later became a well-known
guide at Mount Rainier, yet Muir may have seen in him,
as he called him "small queer Joe," the shadow of that
persecution complex which finally led to self-destruction
in the Tacoma Waterway.

The expedition was on the road before eleven o'clock,

with Muir mounted on a horse called Bob Caribou, Keith on Dexter, a veteran of 1883, and Joe on an unnamed animal which was merely described as "bones." The remainder of the party followed afoot, leading the other animals with their "huge, savage packs," which reminded Muir of a gypsy outfit.

Muir's description of the Nisqually River crossing indicates the route followed was essentially that of 1870, though it is probable that rough wagon roads had replaced the trail for part of the way. Ten miles out they were joined by Van Trump, who came well mounted on a black pony, and evening found them eighteen miles from Yelm, at Indian Henry's farm, where all found beds in the barn. Muir described his host as "a mild-looking, smallish man with three wives, three fields, and horses, oats, wheat and vegetables." It was there at Mashel Prairie that they were afflicted with that sickness which followed them all the way to the mountain, causing them to question the freshness of their canned goods. However, Keith, the only one to escape the malady, says it was caused by rancid butter.[20]

Indisposed or not, they were up at 4:30 on August 10 and soon passing through what Muir called "glorious woods" on a trail as well defended by yellow jackets as ever it had been. An encampment was made that evening at "Forked Creek," a stream which can no longer be identified.

They were on the way again at eight o'clock the following morning, reaching Kernahan's place in Succotash Valley at noon. Warner found the rancher to be a man he had formerly known in Omaha. Perhaps it was the Palisade Ranch Muir was thinking of when he wrote:

> The new comers, building their cabins where the beavers once built theirs, keep a few cows and industriously seek to enlarge their small meadow patches by chopping, girdling and burning the edge of the encircling forest, gnawing like beavers, and scratching for a living among the blackened stumps and logs, regarding the trees as their greatest enemies—a sort of larger pernicious weed immensely difficult to get rid of.

That night they arrived at Longmire's Springs (Soda Springs to some), where the development was yet rudimentary. A few feet north of the soda spring, there was a log cabin about fifteen feet square, with a kitchen shed of cedar slabs, a shake roof and the stick-and-clay chimney so common on the frontier. A bathhouse of cedar slabs stood in the meadow west of the cabin, and the gentle rise behind it had been cleared to allow construction of a hotel, for which the wall frames had already been raised. James Longmire highly recommended the mineral waters to Muir: "Drink at these springs and they will do you good. Every one's got medicine in 'em — a doctor said so — no matter what ails you." It appears those ailing travelers took the advice! It is likely they also listened with interest to the old pioneer's recital of his recent attempt to guide three men to the summit, which was his first and last experience in the guide business.[21]

On the morning of the 12th, "all who had come through the ordeal of Yellow-jackets, ancient meats and medicinal waters with sufficient strength," continued to their destination above Paradise Valley. The trail followed the west side of the Nisqually River to a point about one-half mile above the stream now known as Van Trump Creek, crossed the river, and climbed the ridge on the east side by switchbacks, reaching the top at the shallow notch called Canyon Rim. As the packhorses were plodding up that steep climb from the river, one of them was attacked by yellow jackets. He was loaded with kitchen utensils and photographic equipment, and his bucking sent the tinware flying with a jangling that frightened other horses in the string. Stampfler says:

> Poor Warner was behind. He saw the stuff flying, and began dancing around like a wild man yelling,
> "Stop him; Oh, my plates! my plates!"
> That put the rest of us to laughing and someone yelled back to him,
> "To hell with your plates; it won't hurt them."
> We supposed he was worried about the tin plates, but he meant his photographic plates which he had carried so carefully all the way from Seattle.[22]

Fortunately, none were broken.

From the top of the switchbacks, the trail passed to the

east of the prominent rocky knobs reaching a park-like forest at about 5,000 feet above sea level. Muir observed that "Every one of these parks, great and small, is a garden filled knee-deep with fresh, lovely flowers of every hue, the most luxuriant and the most extravagantly beautiful of all the alpine gardens I ever beheld in all my mountain-top wanderings." He had found his "lower gardens of Eden."

At two o'clock they found a campsite on the east shoulder of the little hill now known as Alta Vista. There, at Camp of the Clouds,[23] at the highest elevation at which horse feed and camp wood were available, the blankets were spread and the kettle was set to boiling, while the weary tourists lounged, sketched, and explored. Muir caught the beauty of that encampment in perfect prose:

> Out of the forest at last there stood the mountain, wholly unveiled, awful in bulk and majesty, filling all the view like a separate, newborn world, yet withal so fine and so beautiful it might well fire the dullest observer to desperate enthusiasm. Long we gazed in silent admiration, buried in tall daisies and anemones by the side of a snowbank.

Warner, the photographer, at once climbed the little hill above the encampment and was entranced by what he saw. He was a vigorous, enthusiastic young man of twenty-four. Originally from Granby, Massachusetts, he came to the Puget Sound country in 1886 as a photographer for Henry Villard's Northern Pacific Railroad. Establishing himself in Seattle, he appears to have been well recommended when Muir needed an energetic cameraman. He was no naturalist, not even an outdoorsman, but he could see the untrammeled beauty around him.

> I looked back and saw such a view as I had never seen before. Oh, if I had the power to describe it. At my feet was a bed of flowers such as I never thought could bloom . . . all was fresh, so new, so clean. The grass was short, fresh and green. There were not any old tin cans, newspapers . . . . I never saw a city park so clean and so nice as that.

There you have a townsman's appreciative view!

That night Mount Rainier stood out in the bright moonlight in such ephemeral beauty that even weary travelers found it hard to go to sleep. The next morning was filled with botanizing, sketching, picture taking and preparation for an ascent of the mountain.

At two o'clock the climbing party, which did not include Keith or Joe Stampfler, left camp, all with light packs except Warner who had decided to take his camera.[24] A steady climb of five and one-half hours over steep snowfields brought them to the foot of the rocky ridge now known as the Cowlitz Cleaver. There, on the saddle between the snowfield they had ascended and the Cowlitz Glacier, Muir saw light pumice sand, indicating a relatively sheltered spot. He recommended that they camp there, at an elevation of 10,100 feet,[25] instead of going higher as Van Trump wished to do. Most of the party were suffering severely from fatigue and nausea and they readily agreed to a halt.

The work of making themselves comfortable for the night was begun at once. Ingraham says:

> Two by two we go to work preparing our beds. This we do by clearing away the loose stones from a space about three by six feet, stirring the sand with our pikes and making a wall of rocks around the cleared place. After a half hour's toil we declare our beds prepared. Hastily partaking of a little chocolate and hardtack, we "turn in," although the hour is early.[26]

One of the party was so sick he lay down apart from the others without making any attempt to protect himself from the cold, and might have suffered exposure if his comrades had not looked after him.

Daniel Waldo Bass, later an associate manager of the Frye Hotel in Seattle, probably did not think much of his accommodations that night. The single blanket brought for bedding failed to keep out the sharp northeast wind, which blew hard enough to shower them with pumice sand all night; but, not being the kind to suffer in silence, he and Warner created a night-long diversion with wise-cracks and buffoonery. Regardless, it was a miserable night in which everyone's sleep was "short and shallow," according to Muir.

In the gray dawn of the morning, those who could wolfed down a little breakfast before the party set off up the mountain, leaving their blankets behind at "Camp Muir" as Ingraham called it.

They followed the rocky cleaver upward, occasionally descending onto the ice of Cowlitz Glacier to pass an obstruction. At the narrow ledge on the face of the great rock which Ingraham would later name "Gibraltar" because of its similarity to the guardian of the Mediterranean, John Hays gave up the ascent and turned back. The snow in the chute was hummocky and gave them good footing, so they were soon up that treacherous groove. At 10:30 a.m. a stop was made at the 12,000 foot elevation to fit caulks into their shoes (half-inch steel caulks and tools to set them had been brought along). At that point, Warner cut his pants off at the knees so he could climb better, depending on a pair of long stockings for warmth.

On the dome there was smooth, hard ice which required great caution in traveling, for a slip would certainly send the unfortunate climber sliding down into one of the many crevasses with which the steep slope was laced. Though a rope was brought, it was not used in the ascent; on dangerous slopes, they had no security other than careful placing of the feet and the support of the alpenstock each one carried. The smaller crevasses were vaulted, but the necessity of crossing the larger ones on snowbridges made the route devious.

Ingraham stayed ahead of the others in the ascent, giving Van Trump the impression he was impatient of guidance and expected to be the only one to reach the summit. Nor did he show any sympathy for the slow members of the party, as Muir did. Piper became exhausted toward the last, and Van Trump "was much pleased by Muir's kindly sympathy for the lad, and with his cheering and encouraging words as he urged the wearied climber to push on to the goal, he meanwhile waiting for him."[27]

The first climber reached the top at 11:45 a.m., August 14, 1888, while the plucky laggard made it an hour later. Without waiting to rest, Warner set up his camera, pulled off his coat to use as a focusing cloth, and proceeded to take the first photographs ever made on the summit of Mount Rainier. He nearly froze his unprotected arms getting those six views (two of groups of climbers atop the icy knob that forms the highest point, one at each crater,

one toward the North Peak and one toward the south peak).
Warner's negatives were unusual for another reason; some
of them were made on sheet film, the first ever used in the
Pacific Northwest. He had received a few pieces in the mail
just before going to the mountain and took them along for
use in case he broke his sensitized glass plates.

Less than two hours were spent on the summit. Van
Trump managed to get one of the party to go with him to
look at the nearly filled-in hole where he, Bayley, and
Longmire spent such a miserable night in 1883. But when
it came to continuing on to the North Peak (Liberty Cap)
with him, none would go.

While they were on the mountain top, clouds began to
form below them, interrupting the grand view and creating
the fear that a storm was brewing, so all but Van Trump
were anxious to be off the summit. Briefly, he stood there,
debating with "the spirit of precaution, reinforced by
thoughts of the wife and the little ones" whether to remain
and attempt the North Peak alone, or descend with his
companions; but he bowed to Muir's opinion of the weather
situation, and, with a "longing, lingering look" at the unexplored peak, he turned to follow the others down.

As is often the case, the descent was more perilous than
the wearying upward struggle. One of the snowbridges
crossed came near to ending the career of Charles Vancouver
Piper. Most of the party had crossed ahead of the twenty-
one-year-old botanist, but instead of walking over, he placed
his alpenstock near the middle and vaulted across. Though
he landed safely on the other side, the bridge collapsed,
carrying his staff down with it. The others offered to lower
Piper into the crevasse on a rope to recover his alpenstock,
but he declined — and lived to become a noted agristologist
of the United States Department of Agriculture.[28]

Van Trump was immediately behind Piper, for he had
been the last to leave the summit, and loss of the snowbridge
forced him to vault over deliberately. He managed it,
landing "somewhat astride of the homeward side of the
crevasse." Muir indicates there was a third mishap when
one of the party slipped and fell. He says: "So steep was
the ice-slope no one could move to help him, but fortu-

nately, keeping his presence of mind, he threw himself on his face and digging his alpenstock into the ice, gradually retarded his motion until he came to rest."

Piper was assisted down the icy portion of the dome with the rope, though the manner in which it was done sounds unsafe in the light of present rope technique. One of the climbers would go down a reasonable distance, set his alpenstock firmly and attach one end of the rope to it with a half-hitch; then Piper would take the other end and walk down past the anchorage as far as the rope would go. However, he did get down safely.

Gibraltar Rock was reached at 3:30 p.m. which assured them a dangerous passage of the ledge, under a bombardment of rocks and ice loosened from the cliff above by the warmth of the sun. Ingraham was narrowly missed by a shower of boulders he described as "singing as merrily as a cannon ball just shot from a thirty-pounder." A different ledge (narrower but less exposed than the one crossed when ascending) was used, yet it was a harrowing experience.

The remainder of the descent was easily accomplished. Below Camp Muir some were able to slide quickly down by means of a standing glissade, balancing with the alpenstock, and the last straggler reached Camp of the Clouds by seven o'clock.

The members of the summit party were easily identified the next day. In addition to the weakness resulting from insufficient nourishment while under great physical strain, there were bruises, sunburned hands and faces, and weakened eyes. Dan Bass was quite sick and so snow-blind he had to be led about the camp, while Warner's sunburned lips became a raw sore.

A warmer campsite was soon found at a lower elevation; then Muir made what are probably the earliest measurements of the rate of flow of the Nisqually Glacier. By setting stakes he found it moved twelve inches in twenty-two hours.

On the 16th, the party began to break up. Van Trump and Dan Bass left for Yelm, while Ingraham, Piper, and Booth passed around the east side of the mountain, across the Cowlitz and White River glaciers to the Bailey Willis trail, which they followed to Wilkeson.[29]

93

On the 18th, the camp equipment was packed and moved to Longmire's Springs, where Muir, Keith, and Warner rested over Sunday. That would appear to be the day Warner took that earliest photograph of James Longmire's development there. The plan had been to start homeward on Monday, but the horses prevented it by returning to the lush meadows at Camp of the Clouds; they had to be tracked and brought back, which consumed the entire day.

A start was finally made at 8:30 on August 21, but trouble with the packs prevented them from getting farther than Kernahan's farm that day. There they were able to get a good meal and a bed in the barn. They reached Indian Henry's farm the next night, enjoying the comfort of his barn for a second time, then pushed on to Yelm, where they arrived in mid-afternoon. At Van Trump's house they had what Warner called "the first good meal we had eaten for two weeks."

So ended a notable expedition to Mount Rainier. From it would come the first description of the mountain and its beautiful alpine parks to reach the American people generally. Though both Kautz and Stevens wrote accounts in the year 1876, theirs were both too early and too much concerned merely with mountain climbing to do much for the mountain. In contrast, John Muir was a popular writer, writing at an opportune time, for the United States Coast Survey had just released the results of the 1870 measurement of Mount Rainier's height by angulation from the west.[30] On the basis of a computed elevation of 14,444 feet, it was thought to be the highest mountain in the United States, which explains the intense interest with which Muir's words were read.

Without a doubt, the period following the opening of the Bailey Willis trail saw many unrecorded attempts to climb Mount Rainier. But there is one which will serve to illustrate the naivete with which such enterprises were often undertaken.

In 1887 a young man by the name of Archibald Alston, Jr., arrived in Tacoma from the East. He soon found work with the engineering corps of the Northern Pacific Railroad, becoming, in due time, a willing instrument of the land grabbing policy established by Bailey Willis a short time earlier. So "Arch" Alston took a timber claim of 160 acres west of Mount Rainier, where he stayed six months in a cabin built jointly with the man on the next claim, after which he returned to Tacoma.

We need not dwell upon the disposition of the title which resulted from that brief residence; of greater interest here, is the attempt made by Alston and his partner, Hershey, to reach the summit of Mount Rainier. He says:

> One day my partner and I started for the top or somewhere near the top of Mt. Ranier [sic]. The second day we got up above timber line (it was the last of August). We crossed one level place with snow on it which looked like a lake, then on up the mountain to another lake of snow. My partner went in up to his arms which scared him so he would not go any further, so we turned back.[31]

There were no fatalities on Mount Rainier that summer, despite the close calls, but tragedy did strike at Longmire's Springs. A small party, including a man by the name of E. H. Hudson and one or two of his sons, walked in by the Yelm trail, camping on "Iron Creek," about where the National Park Inn now stands. The following morning, as they were breaking camp, Mr. Hudson bent over the fire to put it out, and his derringer pistol fell from a vest pocket, discharging as it struck the ground. The bullet made a large wound in his neck, severing an artery.

Maude Longmire was churning butter by the door of the cabin when one of the men ran up to get help. She called her grandmother, who gathered up some bandages and medicines, then went to the camp to do what she could. But Hudson's wound was too serious—he died within an hour. Lacking boards for a coffin, the body was placed in one of the large cedar bathtubs then used at the springs, and burial was accomplished that afternoon beside the trail, a quarter of a mile below the meadow.[32] A low, rock-rimmed, mossy mound remains to mark the grave of the first visitor to die at Mount Rainier.

A week after Ingraham returned home from his successful ascent of Mount Rainier, he and his wife took in a roomer,

95

the Reverend Ernest C. Smith, who had just arrived in Seattle to be pastor of the Unitarian Church. The twenty-four-year-old minister showed such interest in the inevitable recital of Ingraham's mountaineering triumph, that the major agreed to take him to the mountain by way of the Bailey Willis trail. Several weeks later, after the beginning of the rainy season, they were able to make the trip.[33]

It rained as they plodded up the trail to Spray Park, where two inches of snow fell during the night, but the storm passed, leaving the meadows sparkling white in the bright fall sunshine. After some rambles among the still verdant flower fields (Smith was an enthusiastic botanist), they made their way across Carbon Glacier, passed the Winthrop, and moved down White River to the railroad, which took them back to Seattle. It was just the trip to give Smith the "mountain fever."

With the coming of John Muir in 1888, Mount Rainier found a voice it had long needed. His beautiful prose, sparkling with poetic description, excited a public already intrigued with the thought that the majestic peak dominating the Cascades was also the "highest" point in the United States. By presenting the mountain as a thing of beauty, he was able to spread the mountain fever among many who would never dare to climb — he left more eyes turned mountainward than ever were before!

CHAPTER V

WIDENING OF THE RIPPLE

1889-1890

*John Muir's message was opportune in an economic
sense also, for it came just as the Puget Sound country had
attained a measure of stability. The period of vigorous,
lusty development based on speculation, with its preoccu-
pations and urgencies, was giving way to a more ordered
progress; there was increasingly more time and inclination
for the frivolous things of life—among which were excursions into the alpine wonderlands at Mount Rainier.*

In 1889 JAMES KERNAHAN, THE LONELY SETTLER in Succotash
Valley, could say he had a neighbor, for O. D. Allen, a
retired professor of chemistry at Yale University, settled
two miles east of him. That, and the talk of building a
railroad to tap the timber in Mashel Valley, where Van
Eaton had just located, was sufficient hint of the increasing
tempo of events.

The climbing season found "Major" Ingraham eager to
make another ascent of Mount Rainier, and E. C. Smith
just as anxious to go with him. Both were companionable
men, so the party which arrived at Paradise Valley was a
large one, including also Grant Vaughan, a Seattle music
teacher, Dr. L. M. Lessey, "a delightful man to have on
any kind of expedition," Roger S. Green, H. E. Kelsey, and
another Smith with the initials J.V.A., familiarly known
as "Van."

Since the single account written of that first all-Seattle
ascent is now nearly unavailable,[1] the portion concerned
with the climb is reproduced here in its entirety:

On Monday, the 12th of August [1889], a party of seven gentlemen left their camp in Paradise Valley (6,000 feet elevation) with the purpose of ascending Mt. Rainier and spending a night in the crater. Although but three parties had ever succeeded in reaching the summit, we were fully resolved to accomplish our purpose,[2] and, to strengthen our hopes, we had as leader Prof. Ingraham, who had made the ascent a year before. It was nearly noon when we left our pleasant camp, and, turning our faces mountainward, began to climb. We had first to get out of the valley where our tent was pitched. It was 'climb' from the very outset; for the wall of the valley was very steep, and about 1,200 feet in height. This first wall was clothed with short, slippery grass, making the foothold uncertain, so that we very frequently zigzagged in our upward course. Once on the ridge, we headed for the nearest glacier, and were soon following it up toward the central peak. The light reflected from the ice made goggles necessary, so we saw the world through blue spectacles, though not with the traditional effect; for we were an exceedingly jolly party, Having already been in camp several days, walking on a glacier was not a new experience to us; but the day being rather bright and warm, a thousand little streams were coursing over the surface of the ice, and the ice itself was rather soft on top, so we had rather an uncomfortable time of it for a while.

A big ridge which divided two glaciers afforded us an agreeable change at times, in spite of its ruggedness. In the clefts of these bare and weather-beaten rocks, it was a surprise to find bright hued and delicate flowers, sometimes in mats of considerable size. Even at an elevation of 11,000 feet, a few species still gladdened our eyes. Another interesting fact was the presence of insects on the ice-fields; and in one place we even marked a butterfly fluttering about in the sunshine, where there was not even a friendly rock to afford it a resting place.

We pushed on, now travelling over rocky ridges and now following glaciers, until we reached "Camp Ewing" at an elevation of 10,000 feet. Here the last year's party had camped for the night, when they made their ascent.[3] It wasn't much of a camp,—simply a level bit, floored with crumbling pumice, protected from the wind by a cliff on one side and a heap of stones on the other,—but it was the last camping-place we were sure of until the summit was reached. It was not late, however; and some distance before us and looking very near was a mighty rock, like El Capitan of Yosemite, along whose base we knew we must pass next morning,—the most dangerous part of our journey. So we decided to push on and camp as near as possible to its base. Such a climb as we had the next two hours! over the roughest, steepest crags, where each one had to exercise the greatest care not to send loosened fragments down on those who were behind, going into places that seemed inaccessible at first, but where a possible passage opened step by step as we advanced. At last we were close to the base of the great rock (about 2,000 feet high above where we stood), which we afterwards christened "Gibraltar."[4] Here we found a place that offered the possibility of a camp. Possibility, I say, for it was barely that,—a little ledge among the roughest crags, not more than three feet broad by twelve feet long. A hour's vigorous use of alpenstocks succeeded in enlarging the platform so that we could all lie down on it by packing closely, though even then one's feet hung out over the precipice if he extended himself at full length.[5]

Of course, we had enjoyed some views of the nearer scenery in coming up to that height (11,000 feet); but the smoke from forest fires and the clouds had blotted out the distant prospect. Now, however, as sunset drew near, the clouds lowered a little; and like distant islands appeared the white summits of Mt. St. Helens and Mt. Adams. But we were filled with a sense of expectation and waiting; our minds were busy with the morrow's undertaking. Moreover, we were tired with our climb, so, soon after sunset, we rolled ourselves in our blankets, and huddled together for warmth. How heavy the blankets seemed when we were "packing" them at mid-day! How unaccountably light and thin they seemed now! I was awake often during the night, and its spell will long linger in my memory. A smoky summer had made me almost forget what the blue sky was like; and here it was once more above me. The blue seemed deeper and intenser than of old; the stars glittered with the very radiance I had known in New England winters; the moon, though near full, failed to dominate the night as usual; and that camp will always be to our party "the Camp of the Stars."[6] The silence was almost oppressive, broken occasionally by the boom of distant avalanches, telling of mighty unseen forces. Above us towered the perpendicular walls of Gibraltar, below us was the ocean of cloud, and over us the star-studded arch of blue. There was nothing to suggest life and limitations; no houses to shut us in and suggest the hand of man; not even tree or shrub or smallest plant,—only ice, rocks, silence, and starry skies.

Next morning we were on our way betimes, making the dangerous passage along the face of "Gibraltar." For an hour, our way was along a narrow ledge across the face of the cliff, crossing and ascending at the same time, with a perpendicular fall on the lower side, of many hundred feet, to a heap of sharp rocks below. In one place, this ledge was not three feet in width, and was crossed by a sort of gully which made it slope toward the dangerous edge. To add to the awkwardness of the passage, it was very difficult to secure a firm foothold in the loose gravel and unbroken rock. We would set our alpenstocks, and then plant a foot firmly. As we did so, some of the loosened fragments would slide over the edge of the precipice, their rattle on the rocks five hundred feet below warning us what the penalty of a slip would be. This was the most dangerous place, but there were many nearly as appalling. None of us were affected by giddiness, however, and these places did not delay us long. In one place, a few stones rattling down on us from above warned us that that was no place to linger. Indeed, in the afternoon of a warm day, the fall of rocks from above is almost constant. We got out of that place as quickly as possible, taking to the snow as soon as we reached it, in order to get away from the cliff and possible bombardment from above. At eight o'clock, we stood on the top of "Gibraltar," and saw before us the dome of gleaming white, apparently smooth and easy of ascent, though seamed here and there with crevasses that promised trouble to venturesome climbers.

When we had struck out direct for the summit, we found that the snow was anything but smooth, having been blown, or melted, into the form of the roughest kind of billows, and sometimes seeming like slates set up on edge, with hollows three feet deep between them. Progress under such circumstances was not rapid, as you may imagine. The dreaded crevasses, which had formed an insuperable bar in the case of some attempted ascents, did not trouble us greatly. We met with some, of course, great clefts in the ice, fifteen feet wide and a hundred feet deep, some with smooth perpendicular walls, others with masses of icicles so arranged as to form ice palaces of marvellous beauty. But, by following these clefts a quarter of a mile or so, we either reached places where the cleft narrowed so that we could jump it with our alpenstocks or else we found a "bridge." These bridges consist of snow that has drifted across and hardened into a tolerably stable structure. They vary greatly in width and thickness, but usually are pretty thin in the middle, and have to be carefully tested before each one makes the passage.

Three hours of hard climbing brought us to the crater's rim, a little below the summit. Some of the party were badly used up, and cared for nothing but rest; while a few of us at once made for the apex, and explored the crater. The summit was on the opposite side of the crater, a half mile distant in a straight line. We reached it, however, by following the rim of the crater, the bare rocks of which are a hundred feet or more above the snow both without and within the crater. We saw jets of steam issuing from the rocks in several places; but that could wait for an investigation, we thought, and hastened on till we stood on the very apex, 14,444 feet above the sealevel.[7] Our exhilaration had reached its height. Our motto was no longer "Excelsior," but "Eureka."

What shall I say of the view that met our gaze? Though

some clouds lingered in the valleys, they in no wise contracted the limits of our vision. We gazed down on a cyclorama of a hundred miles or more radius,—beautiful, indeed, but so grand as to almost make us forget its beauty. The most prominent features in it were the mountain chains that rose, range behind range, like the serried ranks of a mighty army, with now and then some white-capped monarch overtopping the common herd. In addition to the distant peaks we had seen the night before, our view now included the peaks to the north and east, even taking in Mt. Baker, nearly two hundred miles away. The sensation was entirely different from that experienced on a lesser peak, for there the view is almost invariably shut off in some direction. Here we were far enough above all neighbors to enjoy what seemed a boundless view. The mere vastness was deeply impressive, had there been nothing else to give a spell to the prospect. As our gaze at last came back from the distant horizon and rested on the slopes we had just climbed, how insignificant many of the obstacles that had seemed so great to us a short time before now looked! How beautiful our valley, Paradise Valley, with its graceful groves and brawling stream! How near it seemed! In another direction, we could trace the mighty rivers of ice in their course downward and outward, till at last they came to an abrupt end, and foaming streams appeared in their place. Some of these glaciers took such terrible headlong leaps that their surface was broken, and their substance crushed, till they looked like anything but rivers,—jagged masses and irregular pinnacles of crystal piled in direst confusion, seamed and scarred by chasms cleft in all directions. Others had a more orderly course; and, as the lesser ones united and larger ones were formed, we could see the moraines blackening their surface in regular lines, could distinguish the crevasses following even curves bowed out in the centre, showing that the motion was greatest there. We could trace the spurs of the mountain far down among the magnificent forests, could pick out here and there a crystal lake, here and there a valley we longed to explore.

But even such a prospect could not hold our attention long, when we felt our members congealing under the assaults of the piercing wind that made it difficult even to stand where we were. So we clambered down the crater wall, and, freed from the icy blasts, enjoyed the warm sunshine and the steam. For it is true that from the clefts of the rocks issued, at times, jets of genuine steam, hot enough to give a severe burn. When we wished for water, we filled a tin cup with ice, held it in a steam-jet, and soon had our water. The steam had hollowed out quite extensive chambers in the ice where it came in contact with the crater wall, so that by a vigorous use of our alpenstocks, we soon had comfortable, steam-heated chambers in this strange hotel. We had to be careful, though, not to get too near the

101

steam-jets, as one of our party found to his cost. He had made a couch where he was enveloped in steam.

In the morning, he looked as if he had been in a steam-bath all night, his blanket and clothing positively dripping with water. Of course, he had not gone far from his steam-jet when his clothing froze, and he had to go back into the steam, remaining there until we were quite ready to make the descent. The rest of our party passed a comfortable night without paying any such price for it.[8]

Next morning, we found ourselves on an island just rising above an ocean of cloud that stretched unbroken to the horizon. The sky above was clear and blue, making our isolation in space all the more impressive. But, mindful of the danger in passing 'Gibraltar' after the sun had risen high enough to loosen the stones by melting, we said good-by to the crater, and rapidly made our way toward permanent camp, accomplishing the whole descent in about five hours. When we see the mountain now from Seattle, it is much more to us than it ever was before. We have a delightful sense of companionship with it that can only be born of close acquaintance. Most of our party are anxious to make the trip again next year.

It should be noted that C. V. Piper accompanied that party of 1889 to Mount Rainier, but his harrowing experience of the previous summer had apparently removed all desire to revisit the summit. Instead, he remained at the camp in Paradise Valley botanizing. He and E. C. Smith identified as many plants as they could, sending some plant specimens to the "Young Scientist's" herbarium and some willows to M. S. Bebb, the willow authority at that time.[9] It was the first scientific work on the flora of Mount Rainier and a worthy monument to the party.

As the party led by Major Ingraham was descending from the summit of Mount Rainier, two men were leaving Tacoma determined to reach the top. One was Charles H. Gove, a charter member of the new Oregon Alpine Club, of Portland, and the other was J. Nichols, of whom nothing is known except that he was a resident of Tacoma.

Well-laden with climbing equipment, provisions and blankets, they boarded the train for Yelm, where they obtained packhorses for the trip to the mountain. Their route was the old Yelm trail, along which they traveled until night caught them on the bank of the Nisqually River near Ohop Creek, unable to find the ford. Camping uncomfortably there, they crossed the river early the next morning, making their way to Indian Henry's before feeding themselves or the horses.

Profiting by that experience, they carried hay with them so they might camp where night overtook them without the animals suffering for it. The second day's travel put them past Mashel Mountain, where they camped again on the river's edge. The third day they reached Longmire's Springs.

Gove was impressed by the new hotel built of hand-cut cedar "boards twelve inches wide and fifteen feet long — as perfect as though sawed."[10] But the really fervent testimonial was reserved for the mineral springs themselves. He found them "positively life-giving," and went on to say he had never seen anything to compare with their "stimulating and health-giving qualities," from which we may assume it was still a hard trip in from Yelm and a hot bath was a real lifesaver!

Only one horse, with equipment and provisions for the ascent of the mountain, was taken beyond the springs. The plan was to take the horse to an elevation of 8,000 feet, then leave him near the upper limit of the alpine meadows, while they climbed on; but Paradise Valley proved so unexpectedly beautiful, they decided to stop over a day instead. It was a fortunate decision.

Camp was barely prepared and a kettle of beans put over the fire to boil, when there were signs of a change in weather. The wind veered to the southwest while clouds gradually overspread the sky. On August 18 there were valley clouds, in addition to a thick overcast, with the barometer moving toward storm; so the ascent was again delayed. Gove had been through a summer storm on Mount Hood and prepared accordingly; the tent was restaked, ditched, and the sidewalls were secured by piling rock and wood on them. When the rain came with gale winds at three that afternoon, all was ready — they could crawl into their blankets and out-wait the weather.

The storm raged all night, turning from rain to sleet, then to snow as the temperature dropped to 34°. In the morning, a fire was made with difficulty and they managed some hot coffee; but when the beans, which had simmered intermittently from the time camp was established, were tried, they were found rock-hard and could only be thrown out.

Foreseeing no immediate improvement in the weather, a retreat was made to the relative comfort of James Long-mire's hotel, where they arrived on the afternoon of the 19th, in a "rather moody frame of mind."

The morning of the 22nd dawned clear, so they made an early start, reaching the ridge above Paradise Valley at one o'clock. There the horses were turned loose to graze, packs were shouldered, and the ascent was begun. That evening, they camped in a small, light-weight tent at Camp Muir, finding it a snug shelter after the sidewalls were banked and the damp pumice sand was covered with an oilcloth.[11] As the sunset mantled the peaks of Adams, Hood, and St. Helens in a rose glow, the temperature fell to 28°, and the two climbers withdrew into the first "mountain tent" used on the slopes of Mount Rainier.

After what they called a comfortable night, they were on the way to the summit at 5 a.m., leaving the tent, blankets, and provisions behind. The icy condition of the slopes re-quired much step-cutting, so that their progress was slow. At twelve o'clock they were not quite to the top of the "big rock" (Gibraltar), and another hour passed before they reached the dome. Snow conditions were unusually poor from there to the summit. The surface was broken into pinnacles six to seven feet high and there were many large crevasses. As a result, they were far below the summit at three o'clock when they began to wonder if it were possible to reach the top and return by dark.

At four o'clock a round trip was obviously impossible; so they decided to push on to the crater to spend the night as Stevens and Van Trump had in 1870. The decision was easier for Gove and Nichols, however, for they knew there was shelter in the steam caves, and that the Ingraham party had left some provisions on the summit ten days earlier.

Troubled by obstacles as they were, the route of their ascent was similar to that of Stevens and Van Trump, bringing them out on the summit platform well up on the side of Point Success. From there they crossed the shallow saddle to the crater peak, then climbed its three hundred-foot outer wall to stand on "one of the highest mountains in the United States."

It was August 23, 1889, at 5:10 p.m., and the thermometer stood at 23°. They delayed only long enough to write their names on a card, place it in a box marked "Oregon Alpine Club, Portland," and heap some stones over the container, before seeking shelter. They soon found the can of corned beef, the French chocolate, sardines, and cheese left by their predecessors, taking them and the tin cup (found lashed to a rock near the summit) into a steaming hole in the rocks.

There they cleared away the stones to expose the sandy bottom, took off their outer clothes to keep them dry (they wore two suits), and made themselves comfortable beside the steam jets in their rocky nest. Appetites were poor and fatigue great; so they were soon asleep — to be rudely awakened during the night by cold dashes of wind-driven snow.

The morning dawned clear and cold with a memorable sunrise. Steam soaked clothes were soon changed for the dry ones left outside their shelter and the descent was begun at once. Longmire's Springs were reached at eight o'clock that evening and Tacoma three days later, ending the first ascent of Mount Rainier by a member of a mountaineering club.

Gove was a bold and skillful mountain climber who might have accomplished much, had his personal fate not ruled otherwise.[12] The almost unknown Mr. Nichols gained the distinction of being the first Tacoman to reach the top of the great mountain that Tacoma was already claiming for her own.

Do not overlook a valiant climber who was trudging bravely up the steep snowbanks toward Camp Muir as Gove and Nichols were plodding up the dome. Nine-year-old Christine Van Trump came to the mountain with her parents for an outing, and that day she climbed with her father as far as her strength would allow. It was an adventurous, wonderful day for the child, as the little letter she later wrote to her teacher shows. Fortunately it found a

105

place in the columns of the Tacoma *Every Sunday*. Here is what she wrote:

Yelm, W. T., August 28, 1889.

Dear Miss [Fuller]

Mamma, papa and I were on the mountain. Papa and I went up ten thousand feet, to Camp Ewing. We passed your camp, after putting our names in it. On our way up we went through a beautiful ice tunnel, with green and blue ice in it. We left mamma in Camp Tacoma, near Camp of the Clouds. The next time I go up to the mountain I am going to the top. While we were at Camp Ewing we saw two men climbing the crater. They were so small that they looked like moving black spots in the snow.

Christine Louise Van Trump.[13]

Christine's optimistic wish to reach the summit of Mount Rainier was to go unfulfilled; though of plucky spirit, she had a body crippled by the nervous disorder known as "St. Vitus dance." Remember, as you look at Christine Falls, that they are named for a little girl who made one of the really great conquests of the mountain, though she got but two-thirds up!

Once set in motion, the process of settling up a frontier goes inexorably on to its destined end. The summer of 1890 opened with a new homestead on the Yelm trail, at a place which would one day be known as Elbe, while the trail itself was a little less important by the completion of grading on the wagon road from Yelm to Mashel River.

The climbing season apparently opened early with what is reputed to have been a solo climb of the mountain. Unfortunately, the evidence is meager and not very convincing, and it lacks confirmation from any independent source.

Oscar Brown not only climbed alone but said nothing about his feat until two years afterward, when he had gained some prominence by a publicity stunt. That first interview appears lost beyond recovery with the little newspaper it was published in, but if he was no more definite than in a later interview with a *Tacomian* reporter, very little was lost.[14] Since Oscar Brown's purported 1890 climb is not very convincing, it is only fair to repeat the interview as published.

Oscar Brown, of Enumclaw, ascended the mountain in 1890 and 1891. In an interview with Mr. Brown in regard to his first trip published in the Enumclaw *Evergreen* of January 15, 1892, he said:

"Standing in the center of the basin which is the apex of the three principal peaks of the great mountain, the highest peak, the one containing the craters, lies to the North-northeast while the other two lie to the West and southeast. This basin is triangular in shape, and the moving glaciers are continually breaking and falling over the precipices on the three sides, thus giving rise to the White, Carbon, Muddy Cowlitz, Nisqually, and the North and South Forks of the Puyallup rivers. The two forks which go to make the White River have their sources in the northeast and north portions of the mountain; the Nisqually has its source in the southern part; the Carbon in the northwest; the Muddy fork of the Cowlitz in the southeast; and the North and South forks of the Puyallup River in the west and south-west.

"While here on this trip, I was seized with a determination to ascend to the highest point, if such an ascent were possible, and strapping my luggage to my shoulders commenced to climb. The work was very difficult, tiresome and often times exceedingly perilous. Often without warning, great icicles, 8 and 10 ft. long, would break from their elevation and come tumbling down, dislodging the rocks as they came. They were not very agreeable visitors to those below. But there were other disagreeable features. The cold is intense, and the wind howls around the rocks very unpleasantly. There is also danger of snow-blindness. But my mind was entirely made up to accomplish the ascent if possible, and after hard work I was at last standing on the highest point of Mt. Tahoma, 14,444 ft. above the level of the sea.[15] Close investigation proved that two craters exist on this peak, though one of the oldest, now apparently entirely dead is but partially to be seen, at the time of my visit being almost entirely covered with the deposits made by the eruptions of the new crater, and several feet of snow on top of this. It lies generally to the southwest of the newer one, which has to some extent though incorrectly been described before. The center of the latter crater was covered with snow and ice at the time of my visit. While around the interior edge all was warm, and slight volumes of steam were continually arising therefrom. The volcano has doubtless been active from this crater less than 100 years ago, and from all appearances may break out e're another century rolls around.

"The diameter of this crater has been described to be three-

107

quarters of a mile, but this is not so. The air being very light, the vision is greatly deceived as to distances, and when standing on one edge of the crater the other edge seems to be but a very few yards away. Wishing to satisfy myself in regard to its area, I started to 'step it' the shortest way. After walking for 170 yards I broke through the covering of ice, and fearing I knew not what, immediately changed my course. At this point the steam from below had warmed the rocks and melted the ice and snow, leaving but a thin crust at the top. As I broke through my feet struck rocks underneath, and I easily escaped. I was not any farther than thirty yards from the other edge, and I am ready to state that the area of this crater does not exceed 200 x 300 yards.[16]

"The descent is more perilous than the ascent, as one encounters greater difficulties. I always carried a dirk knife, which greatly assists in climbing on the snow and ice. All the way down I found volcanic rocks and cinders in great profusion, some resting on tree stumps less than a century old, proving that the volcano was active not many years ago; one variety of cinder is very spongy, and so light that it will float on water sometime before it sinks. This variety is found, carload after carload, wherever it could find lodgement away from the winds. In fact, I have found them on the summit of the Cascades twelve miles away."

Mr. Brown said it was next to impossible to make the ascent from any other direction than the way by Longmire's springs. "I have made many attempts and failed. At one time in an attempt to climb the northwest side, I was forced to cut 1300 steps straight up, where, if I had made one slip, I would have fallen hundreds of feet below. It is often necessary to leap five or six feet across chasms, where stones might be dropped and fall hundreds of feet; and here the knife comes handy to strike in the frozen snow and assist one to hold his position. At one time while walking up a narrow incline with a fellow traveler I heard the warning sound of a great rock crashing toward us, and we had but just reached a place of safety as the huge stone whisked past the place where we stood."[17]

And so ends a potpourri of geological, geographical and mountain climbing rubbish. Most significant is that fact that Fay Fuller, who had a nose for mountaineering news (and an editor-father to print it) did not record Brown's ascent in the articles she wrote that summer. We will meet Oscar Brown again, where he is better supported, but for now, he is dismissed.

Not quite a month after Oscar Brown claims to have

reached the summit of Mount Rainier, the principal party of the season gathered for a climb. Gathered is the right word, for it was made up of members of three parties encamped at Paradise Park.

Earlier that summer Miss Fay Fuller, the young school teacher of Yelm and social editor of the Tacoma *Every Sunday* (her father's newspaper), accepted the invitation of Mr. and Mrs. Van Trump to accompany them and their ten-year-old daughter Christine on a camping trip to Mount Rainier. Fay had some hopes of climbing to the summit, and was dressed for it, though we might not think so now. Her outfit consisted of "a thick blue flannel bloomer suit, heavy calfskin boy's shoes, loose blouse waist with innumerable pockets in the lining, small straw hat";[18] nothing very *chic*, for sure, yet it was considered *very* immodest.

On August 4 they left Yelm horseback, with extra clothing in their saddlebags, and tent, blankets, and provisions on a packhorse. Crossing the Nisqually River bridge, they followed the new wagon road, passing houses every few miles to Ohop Creek. The trail down to the stream from the high bluff on which the road ended, was so steep it was necessary to dismount and drive the horses down. Late that afternoon, seven and one-half hours after leaving Yelm, they arrived at Indian Henry's where supper was soon made over an open fire and beds were fixed on new hay in the barn.

Just at dark, two other parties came in. One, led by E. C. Smith, included Miss Katie Smith, Miss Tillie J. Piper, Dr. J. J. Sturgis, Robert R. Parrish, and W. O. Amsden, a Seattle photographer;[19] the other was a Longmire family group, consisting of Elcaine Longmire, his sons Leonard and Washington, his daughter Maude, and her guest, Miss Edith Corbett, of Des Moines, Iowa. It was a large and merry gathering.

The camp began to stir at four in the morning, and the Van Trump party left at 7 a.m., without the others, who were not yet ready. They pushed on along the shadowy forest trail around Mashel Mountain, then up the broad valley of the Nisqually River, Van Trump occasionally raising the old cry of "Hyack-sojers!" to gallop his little cavalcade past warring yellow jackets. By making a fourteen-hour day of it, they reached the Palisade Ranch that evening.

Kernahan had, by then, "an excellent farm with about forty acres of cleared land, a comfortable house, a young orchard, and a well-cultivated garden"; Miss Fuller thought it a delightful place to visit.

The third day's ride took them through to Camp of the Clouds, a safer trip since James Longmire had built the bridge over the Nisqually at the foot of the "switchbacks." Camp was set up handy to wood, water, and grazing; then there was a day of gathering and pressing flowers before the Seattle and Yelm parties came in.

They had not traveled as swiftly. After leaving Indian Henry's, they made the usual camp on Nisqually River above Mashel Mountain, spending the third night at the "Springs," after a near tragedy while fording the turbulent Rainier Fork (Tahoma Creek). At that crossing, one of the young ladies of Smith's party was thrown from her horse and washed some distance downstream before she was pulled unconscious from the water.[20]

The Seattle and Yelm parties did not stop at Camp of the Clouds, but passed on to set up their camps at a slightly higher elevation on Paradise Hill. There were already several recognized campsites: at the lower end of Paradise Valley, beside a little pond, was "Camp Tacoma," named by the Dodge party in 1889; of course, John Muir's "Camp of the Clouds" on the slope of Alta Vista remained popular; and the "Paradise Camp," on the little hill at the head of Paradise Valley was becoming so. Subsequent years would see the campsites and the names multiplied and shifted until all lost their identity in the confusion.

The next day Miss Fuller went over to the other camp to visit, finding several there who were planning an ascent. Upon expressing her hope of reaching the summit, she was invited to join them.

Much of the forenoon of August 9 was spent preparing for the ascent; appropriate clothes were put on, worn shoe caulks were replaced and blanket rolls were made up. Eleven o'clock found three of the Seattle party and Miss

110

Fuller ready, so they started under the leadership of E. C. Smith. Each had rations for three days rolled in a pair of blankets, which were carried slung over one shoulder with the ends tied together at the opposite hip. In addition, Smith had a three-foot mercurial barometer, prismatic compass, hatchet, and fifty feet of rope; Amsden had his heavy eight by ten view camera, and Parrish had a field glass.[21] All had alpenstocks, goggles, dark veils, and gloves.

They toiled up the rocks and over the snowfields—passed Plummer's camp on the exposed horn of Anvil Rock, passed Camp Ewing nestled in slide rock, and on up the seemingly endless snow slope toward Camp Muir. There they rested an hour while waiting for the climbers of the Yelm party to come up.

Elcaine Longmire, with Maude, Leonard, and Washington Longmire, and Miss Corbett, reached Camp Muir at six o'clock, too late for the climb to continue on to Camp of the Stars, as Smith had wished. After a meal of corned beef sandwiches, cheese, and sticky chocolate, beds were prepared on the loose pumice. Elcaine had carried up a small tent instead of blankets, and it was soon raised on alpenstocks to give welcome protection from the chill of evening.

Just before reaching Camp Muir, Amsden broke the ground-glass of his camera while putting on his pack after a rest stop. Fortunately, a fragment of glass large enough for focusing remained, allowing him to photograph the impressive scene below, where the shadow-filled valleys lay overtopped by sunlighted peaks. Smith was not so fortunate. He broke the delicate barometer in the process of adjusting it for an observation. Putting the fractured tube away, he merely remarked that the accident had changed his trip from business to pleasure!

At nine o'clock that evening a torch was waved, and three reports from a gun out of one of the camps near Paradise Valley assured them it was seen. Then all the climbers retired to the tent, except Mr. Parrish, who made himself windproof with an oilcloth in a nest among the rocks.

The cramped, cold, wakeful night ended at 4:30 a.m.;

111

then blankets were rolled and the ascent was begun. Breakfast was limited to some chocolate nibbled as they climbed, but even that was too much for Miss Corbett, who became sick soon after starting. Seeing that her illness was likely to delay his party until too late for a safe passage around the face of Gibraltar, Len Longmire joined Smith's party, which thereby came to represent Seattle, Tacoma, and Yelm.

It was tedious climbing for all on that ragged, sun-scorched ridge leading to the crucial "ledge," but particularly for Miss Fuller in her hot and cumbrous bloomer suit. Yet they dared not stop to rest for they were racing the westward sweep of the sun, which would make the narrow trail below the great cliff frightful with a rain of debris melted loose from the heights.

The rope was used in crossing the ledge, though only as a hand line; yes, Miss Fuller even made the now ridiculous sounding statement that "it would not be wise to tie the party together." In its implication that the individual should not imperil the party, how different the attitude is from the present recognition that the party should safeguard the individual. While crossing, Mr. Parrish contributed his pack to Nisqually Glacier by a bad pitch, or catch, as it was passed to another member of the party.

The steep chute was icy, forcing the leader to cut steps with the hatchet for fifty feet. While he worked, Smith was hit twice by bounding rocks, but managed to keep his balance, to the great relief of the others moving up from one insecure niche to the next in an agonizingly slow procession. Upon better footing at last, the top of Gibraltar was reached about noon, after seven hours of continuous climbing.

There a half-hour halt was made to lunch by a little trickle of melt water which served to fill the canteen they had brought empty from Camp Muir. It was an all too brief rest, sheltered from the gale winds already blowing, before pushing on.

The route led through a parapeted aisle between the heads of the Cowlitz and Nisqually glaciers, then upward across an area of sun cups where the surface resembled frozen billows with crests three feet high. A shortness of breath soon developed to limit their progress to a few feet between rests; it was a case of "quickly exhausted, soon refreshed" as Miss Fuller said. Several crevasses were crossed on snowbridges by the now roped-in party before what was called a "blind" crevasse brought them to a halt (apparently an offset crevasse with the upper lip jutting over the lower, from the description). After much probing with his alpen-stock, Smith found a safe crossing and the climb was con-tinued to the rim of the large crater, where they arrived at 4:10 p.m. in a bitterly cold gale of wind.

Almost afraid something would keep them from the top, they hurried on to stand upon the knob of ice which is the summit. It was 4:30 p.m., August 10, 1890, and the first woman had climbed Mount Rainier. For a little while she stood there with the others viewing the partly cloud-covered panorama and forming those vivid impressions which later reappeared in her newspaper articles. But the violent wind soon drove them to the shelter of the large crater where they steamed themselves warm before taking shelter for the night in an ice cave on the eastern side.

Mr. Amsden took no pictures, as he felt unwell, and he was soon joined in misery by Miss Fuller, who fell prey to the nausea of altitude sickness. Despite a lack of interest in food, Smith melted some ice in a cup held over one of the steam vents and soon had hot beef soup ready for those who could get it down. After rubbing their feet with whisky, the climbers settled down to a comfortless night in their much too light blankets.

Miss Fuller says she was the only one to sleep, but even she had wakeful spells in which to watch the meteors overhead and listen to "God's music"— the rumbling of avalanches, and the sound of the wind's passage, first softly as a running brook then with the roar of heavy surf. Toward morning a light snow fell, bringing with it cold which froze the mustaches of the men and the shoes of all. Only Smith made the effort to view the five o'clock sunrise which was powerless to temper the 18° gale sweeping over the summit.

An immediate descent seemed wise, so they started down at 6:30 a.m., after a hurried breakfast (dried prunes for Miss Fuller). A triple notice of their presence was left behind them and some mats of crater moss were taken as souvenirs in lieu of photographs, for the dark slide was frozen fast in Amsden's camera, rendering it useless.[22] The weary, blanket-wrapped figures descended slowly through clouds, buffeted by a never-ending gale which occasionally took a member of the party off his feet.

The steps were gone from the chute above Gibraltar ledge and had to be re-cut, making the descent seem even more dangerous than the upward climb had been. Miss Fuller says they "had no ambition to hurry."

At Camp Muir they met the Hitchcock-Knight-Watson party, bound on what has been facetiously called an editorial ascent, gave them such helpful information as they could, and then passed down the mountain to the friends awaiting them above Paradise Valley.

The following day, sunburn laid Miss Fuller low, despite the heroic measures she took to avoid it, and Smith who was also a victim, though he didn't let it keep him from crossing Cowlitz Glacier with Amsden on an exploring trip. On this first of many trips up the mountain, Len Longmire showed the hardihood that would later stand him in good stead as a summit guide — he climbed in summer clothes and wore no gloves, yet suffered no ill effects!

According to Miss Fuller, there were twenty-six people encamped in Paradise Park at the time she made her ascent, and the first "Mountain Notes" to appear in the Tacoma *Every Sunday* give some idea how those who did not climb amused themselves. The Van Trumps made what probably was the first ice cream eaten at the mountain, a Mr. Alling built a horse sleigh for use on the snowfields, and, sadly, "Jim Crow and party killed ten large goats, several kids, and considerable other game."

Breaking camp on the 16th of August, the Van Trump party returned to Yelm, where Fay Fuller wrote a detailed account of the climb for her father's newspaper. The fact that a woman had climbed what was considered the highest mountain in the country on an equal footing with men was big news — and a strike for women's rights, as Clara Berwick Colby bothered to point out;[23] but it was not Miss Fuller's clarion call, "I expect to have my example followed

114

by a good many women," that is the measure of the importance of her 1890 ascent. Its real value lay in heightening her interest in mountain climbing to the point where she began that series of feature articles in *Every Sunday* and *Tacomian*—articles which would bring the mountain fever to epidemic proportions. In Fay Fuller mountain climbing found an effective voice.

The Smith party returned to Seattle about the same time, and we may presume Mr. Parrish went back to Portland—to his writing and the loneliness which eventually brought him to his death;[24] while Amsden probably went into his darkroom to produce that series of Mount Rainier photographs, of which Smith wrote: "Amsden's photographs have a historic interest. Without them there would have been no illustrated lecture in Washington at the crucial time for the proposed park."[25] To which it might be added that the photographs were no more important than the presence of Reverend Smith to give that lecture.

Remember the party pushing upward as the successful Smith-Fuller climbers descended to Camp Muir? They provide just the anticlimax with which to close the season of 1890 — a lucky ascent by novices!

Like the three gentlemen from Snohomish, Solomon C. Hitchcock and Arthur F. Knight (of the Tacoma firm of Hitchcock and Knight) were unaware that Mount Rainier had been climbed. Just what induced them to have a go at the mountain will probably never be known; anyhow, Hitchcock interested his brother Will, a printer of the Fairbank, Iowa, *View,* and he came West for a try, bringing with him F. Van Watson, a reporter of the Dubuque, Iowa, *Daily Times.* It was quite a party; the first two knew nothing about mountain climbing and the midwesterners knew nothing of mountains, not even by sight.

Preparations for the ascent can be passed over without comment, with one exception. An aneroid barometer was borrowed from the firm of Anderson and Borrenger, jewelers and opticians, doing business in the old Tacoma Theatre building. The climbers were to keep a daily record of observations and Mr. Borrenger was to do likewise at the store. Thus, they would have a basis of correcting their observation of the height of Mount Rainier's summit — theoretically, at least!

Since this ascent has received scant attention in print, the following excerpts from Knight's diary will present it best:

> We took the train to Yelm Prairie August 7, 1890. Tickets $1.25 each. Could not get a horse there so started to pack our outfit. Had about 50 lbs. each, went twelve miles the first day and camped by a small creek.
>
> August 8, got up early and went about two miles further and bought a horse for $40.00 which made the going better after that. About fifteen miles further on we camped beside a small spring.
>
> August 9. Nothing eventful today, went twenty-two miles through heavy timbe[r], saw many trees that would scale 20,000 feet or more, eight to ten feet through. Camped beside a roaring mountain torent.
>
> August 10. Crossed the river at the ford and made the base of the Mountain at 10 o'clock and had dinner at Longmire's.
>
> Started on our long march at 1 o'clock reaching Paradise Park at 7 p.m. Had supper and set up camp. Visited Camp of the Clouds. Met Mr. Van Trump an old mountain climber who gave us some good instructions as he had been to the top three times.
>
> August 11. Made an early start and went up to 11,000 feet. Met a party coming down who also gave us some useful information. This party consisted of Dr. Smith, Mr. Anderson,[26] Mr. Longmire, Mr. Parrish and last but not least Miss Fay Fuller, the only woman who had ever reached the summit.
>
> We then started on the most perilous journey we had ever made. At 7 p.m. we dug out a crevice in the glacier and camped there overnight.[27] Used our shoes for a pillow and when we went to put them on they were frozen stiff. We had a bottle of Jamaica Ginger, which the four of us drank raw and felt better instead of worse.[28]
>
> August 12. Took an early start at 4:30 A.M. and started up the glacier. Had a very perilous trip but reached the crater at 8 A.M., tired and half sick. Picked up specimens from the tip top and brought them home with us. Had a small prospecting pick we used to chop steps in the glacier coming up and with this we picked our initials in the face of a huge boulder on the rim of the crater. I have often wondered if they were still there.[29] We got water from the condensed steam, hot enough to make beef tea. We went down under the ice in the crater about 50 or 60 feet. The sulphur fumes made us all sick. Our faces and

116

eyes became swollen. We stayed in the crater as the snow was too soft to come down that day.

August 13: Got up at 4:30, broke camp and started back at 5:30. After a perilous trip reached Paradise Camp at 12:30 had a good dinner, went from there to Camp of the Clouds where we rested the rest of the day. At the camp we met Mr. Van Trump and wife, Miss Fay Fuller, Mr. Vincent and Mr. Murphy.

August 14: Explored the glacier of the Nesqualie River. It was ten miles long and we only went over a part of it. Had a very interesting trip and got back about noon, had dinner and hunted up the horse, preparing to start home the next morning.

August 15: Started for home at 8 o'clock reached the Springs at 11 o'clock. The trip to Tacoma was uneventful, arriving there August 18, 1890.[30]

So much for the boys who ate the sandwich Fay Fuller left in the ice cave. Theirs was the highest "high camp" used on Mount Rainier, and they were the first climbers to spend two nights on the summit; for novices, traveling on gingersnaps and brandy, that is doing very well.

The barometer? The observations on the mountain were carefully made, but when the borrowed instrument was taken back to Mr. Borrenger, they found *he* hadn't kept a record of the sea level readings for he had no idea they would ever get to the top. So much for the "best laid plans."

A. F. Knight returned to Mount Rainier several years later, and will be met again. As for the others, Sol Hitchcock remained a Tacoma resident until his death; Will Hitchcock moved to Sunnyside, where he became the editor of a little newspaper; and Van Watson went back to Iowa and life as a school teacher.

At least thirty-eight climbers had reached the summit of Mount Rainier by the end of the 1890 season, and of that number only two were from Tacoma. The townspeople were undeniably slow to take up mountaineering, yet that is not a measure of their interest in the mountain. The Northern Pacific Railroad's persistent use of the name "Mount Tacoma" after 1883, in order to identify Mount Rainier with its western terminus, was popular in the city, and the local newspapers developed that urban pride into an exaggerated possessiveness which finally became a crusade.

117

Inevitably, the United States Board of Geographic Names was asked to change officially the name of Mount Rainier to Mount Tacoma. At a hearing granted in 1890, the Board carefully considered the matter, and, "without a dissenting vote reaffirmed the name Rainier given it by Vancouver when he first saw the mountain."[31] That action, by a dignified body of men representing scientific and professional services of the government, such as the Smithsonian Institution, Coast and Geodetic Survey, Navy Hydrographic Office, Lighthouse Board, Corps of Engineers, and the Post Office and State departments, should have been utterly final.

The Northern Pacific Railroad did acquiesce in the decision, as a delegation of citizens soon found when they called upon Charles S. Fee, the general passenger agent. He is reported as saying to them, "Gentlemen, we have carried this farce as far as we are going to for advertising purposes. The name has been officially declared to be Rainier and that is what we shall call it. You can call it what you please."[32] Yet the Tacomans were not the least daunted by the loss of their powerful ally — agitation to change the name of Mount Rainier was stubbornly continued for thirty-five years.

In Fay Fuller, Mount Rainier found another effective voice. Her newspaper columns began that series of "Mountain Stories" which, in time, became such a unifying force, stimulating interest in mountaineering, drawing mountaineers together, and creating a broad base of interest upon which a popular movement to obtain a national park at Mount Rainier was later founded.

CHAPTER VI

PEARL OF GREAT PRICE

1891-1892

*The publicity which followed Fay Fuller's ascent of
Mount Rainier heightened the slowly developing interest
in the mountain, leading to a welter of largely unrelated
events without apparent direction. And yet, it was from
among those confused happenings that an altruistic proposal emerged to shape the mountain's future.*

NOT LONG AFTER HER ASCENT of Mount Rainier, Fay Fuller
joined a group of enthusiastic Seattle climbers in forming
the first mountaineering organization in Washington State
— the Washington Alpine Club. Its existence is attested by
the following small program announcement pasted in a
scrapbook[1] now in the possession of the Mazamas:

Washington Alpine Club

ENTERTAINMENT

G.A.R. HALL, HINCKLEY BLOCK, SEATTLE

Saturday Evening, April 11, 1891

PROGRAMME

A Trip to Mt. Baker.......................E. S. Ingraham
In the Olympics.............................C. V. Piper
"The Mountain"...........................Miss Fay Fuller
The Purpose of the W.A.C................Eugene Bresacker

Compliments of———

The club appears to have had broad objectives, judging
from the topics discussed at that meeting, yet it seems to

119

have been an immediate and complete failure, for it disappeared without further mention in the available records of that day. It would be interesting to know the fate of the state's first organization of mountain climbers, but that is probably lost along with the early files of Tacoma *Every Sunday*.

Leaving the first Washington Alpine Club to fade into oblivion, the summer of 1891 began with some buffoonery centering upon a publicity stunt.

North of Mount Rainier, where White River broke out of the mountains, lay the hopeful townsite of Enumclaw, which then had a population of 125, and "the greatest prospects for soon becoming one of the foremost cities of the state." It also had a little newspaper, the *Evergeen*, published by A. G. Rogers, who was proprietor, editor, printer, and devil — in fact the entire staff. As a matter of course, that weekly took a leading part in promoting the town and its interests.

From the vantage point of time and a humorous disposition, Mr. Rogers says:

> Each week the *Evergreen* told of the wonderful possibilities of fortune to would-be investors in our town property; and we really felt quite important. Like those of many other promising townsites, we began to feel that advertisement was necessary, and the plan that ripened into reality was the planting of an Enumclaw flag on the summit of Mount Rainier.[2]

Now the town had a mountain climber of considerable local reputation in a citizen by the name of Oscar Brown (who was also owner of several hundred acres of the best land in the townsite) . He was generally considered to know "every foot of ground within a radius of ten miles of Tacoma's peaks," and so was the logical one to lead the numerous volunteers. Yet the eve of departure found only three of those adventurers really willing to go with him: A. G. Rogers, P. L. Markey, and Silas Balsley.

The climbing party left for the mountain on June 22, after a public send-off at which they were presented a flag purchased by the men of the town, and a pennant made by the women, who had embroidered on it the word "Enumclaw."[3] In a holiday mood, they passed through Oceola, with cheers, then Buckley, where the *Banner* men stopped them for an interview ("We told them all we knew; it didn't take long"), arriving at Wilkeson that night. There they dispatched a chamber of commerce sort of communique to the *Evergreen,* before setting out on the old Bailey Willis trail.[4] Southward they traveled through ancient forests, dipping into the valley of the Carbon River, crossing the Mowich below Palace Camp and on across Puyallup Valley to the ridges which cradle the headwaters of Mashel River, then down into Nisqually Valley to a junction with the Yelm trail near present Ashford. The 26th found them at Longmire's Springs, from which another optimistic dispatch was sent the *Evergreen,* despite the rainy, miserable weather.[5]

Paradise Park was reached two days later. There a base camp was dug out of snow blanketing the high meadows, the horses were turned loose to graze in a nearby valley, and the climbers settled down in "Camp Lakeview" to await favorable weather.

The 2nd of July was the day chosen for the ascent. Starting at 3:40 a.m., laden with extra clothing, a camera, and a twenty-three-foot flagpole cut from an Alaska yellow cedar, they toiled upward, reaching the foot of Gibraltar by 10:30, the top of the rock shortly after noon, and the rim of the east crater at 2:30 p.m., a "completely worn out set of mountain climbers." From there they passed around the rim to the icy summit knoll, which was found unsuitable for planting the flagpole. A better spot was found thirty feet below, and there they were able to pick out a hole two feet deep through loosely consolidated rock in which to erect the staff with the flag flying at the top above the Enumclaw pennant.[6]

A photograph of the flag and several of the crater were made and some time was spent in exploration; then, after two hours on top of the mountain, the descent was begun. They considered the downward trip the more dangerous, particularly under Gibraltar, but reached the base camp safely at 11:00 p.m., after nineteen hours of almost constant exertion.

The well used-up climbers remained several days at

121

Camp of the Clouds where they were treated to a Fourth of July snowstorm while recuperating from fatigue and sunburn. The trip home was prolonged to the 30th by hunting and fishing. On returning, they were disappointed to learn that the flag had not been seen from Enumclaw, or from any other town; it probably was destroyed by violent winds before anyone below knew it had been successfully raised! In its immediate consequences, the ascent was a failure, for the meagre publicity was no help to Enumclaw, or the *Evergreen,* which Rogers wryly remarks did not remain "evergreen."

Another party climbed Mount Rainier during July of 1891, to descend with a tale of wonder and inadvertently bring a disquieting fact to the public notice.

A party consisting of Dr. E. A. Stafford, V. R. Tucker, H. Paulson, and S. E. Glinn left Sumner on July 22 for an outing at the mountain. Three days later they reached Paradise where they met a group of Longmires and their guests up from Soda Springs, and a large Seattle party which had come in by way of Yelm. Accompanying Elcaine Longmire was his son, Leonard, his daughter, Susan, Edith Corbett (who missed sharing Fay Fuller's honors in 1890 because of a "queezy" stomach) , and Professor Allen's son, Ed (mistakenly called "Mr. Elm" in a Seattle *Post Intelligencer* article).[7]

While encamped in the alpine meadows, the visitors saw a column of smoke "reaching thousands of feet in the air," over the mountainous country twenty-five miles to the southeast. Though a large forest fire could hardly have been a novelty to any Puget Sounder of that day, the towering mushroom of smoke, billowing upward in changing volumes and colors according to the fortune of the fire below, soon led them to imagine a new-born volcano, which they identified as a non-existent "Chimney Peak."

On the 29th, the Yelm party, accompanied by two of the Sumner tourists, Paulson and Stafford, and Miller Cooper of the Seattle party, began the ascent of the mountain under the leadership of Len Longmire.[8] The climb was made by way of Gibraltar with the first night spent in the vicinity of Camp Muir and the second in the east crater. While the climbers were toiling upward, the great fire near the crest of the Cascades continued its awesome destruction of the ancient forests. As the red glow of the creeping fire-front became visible from the heights, its resemblance to a flow of molten lava strengthened the impression of an active volcano.

A search was made for the flag placed on the summit by the Oscar Brown party earlier in the month, but only the rocks mounded around the shattered butt of the staff could be found, all else having been carried away.[9] The descent was begun on the morning of July 31st, proceeding without difficulty until the Gibraltar ledge was reached; there the usual rock-fall nearly brought tragedy when a boulder passed so close to one of the young ladies that it brushed her skirt.

While not an unusual ascent, this one deserves more notice than it has received. Two more young women had followed Fay Fuller's footsteps to the summit. Len Longmire had begun his more than twenty years of guiding on the mountain,[10] and Dr. Stafford's report of an active volcano stirred up a lively scientific controversy. But that is not all; the Sumner party brought back some hunting trophies, bragging enough to sting the public conscience for the first time. A Tacoma editor put it this way: "During their expedition the party saw a great many goats and as they brought several skins home it is safe to conclude that during their trip they somewhat lost sight of the state goat law." While such abuses at the mountain would continue, and grow worse before better, a step in the right direction was taken with that remark.

Perhaps the Gibraltar ledge *was* more treacherous in the summer of 1890 than it had been previously, and again, it might only have seemed so. Anyhow, such a comment was made by Fay Fuller in her article in *Every Sunday* and Van Trump seems to have endorsed it.[11] But his choice of the west side of the mountain for an alternative route was probably dictated more by his desire to reach the unconquered North Peak (Liberty Cap), which he had concluded could not be reached from the summit after the rigorous climb by the Gibraltar route. To be first upon the remaining summit of Mount Rainier, as he had been upon the others, was his obsession. And so, as the summer of 1891 came on, he laid plans for an attempt from the west on the peak from which he had been thrice barred. Just how he picked his companions for the ascent will probably never be known, but the choice was a poor one, as will be seen. Out of it developed a minor drama complete with hero, villain, and suspense.

Late July found P. B. Van Trump pressing mountain-ward with two companions. One was Doctor Warren Riley, the elected health officer of Olympia,[12] who had once served under Kautz in the cavalry,[13] and the other was Almond Drewry, the eldest son of a Thurston County pioneer of 1853.[14] They took saddle horses, two packhorses, provisions for a month's stay at Mount Rainier, hunting and climbing equipment, and Riley's deerhound.

The extent of the changes taking place along the route to the mountain are apparent in Van Trump's description of the trip. At Eatonville, where they stopped the first night out of Yelm, they found a general store, a restaurant and lodging house, a blacksmith shop, a shoe cobbler's shop, a feed stable, saloon, and a post office with weekly mail service.[15] The second day's journey took them past the cabins of timber claimants and the homesteads of some of the 300 settlers who formed the population of the timbered valleys east of Nisqually Plains, as well as three log cabin "stores."

From their second stopover at Kernahan's ranch, the party followed the Yelm trail to Rainier Fork (Tahoma Creek), where they took to the dim trail used by Indian Henry on his goat hunting excursions to the parks which now bear his name. It followed the west bank of the stream for two miles, then crossed over and climbed steeply up the high ridge on the east to break suddenly through a rocky gap into the flower covered parks with their islands of gracefully spired trees backgrounded by the great mountain which seemed "to pierce the blue vault of heaven itself with its triple summits."

Within the park, near a cold mountain spring, the wall tent was set up and the horses were turned out to graze. Riley and Drewry began hunting, soon killing four goats

124

and some lesser game, but their sport, as well as the immediate prospect for an ascent of the mountain, was delayed by an illness of Drewry's and bad weather. Snow began to fall on the fifth, continuing so heavily on the following day that it was necessary to twice shake the accumulation from the sagging tent canvas. Not only did the storm make camp life uncomfortable, but it also laid up misery and the basis of ultimate defeat in the deep layer of new snow deposited upon the upper slopes of the mountain.

The ascent from the west began on the morning of August 10, 1891. Each of the climbers carried a double blanket, two days' provisions, a bundle of dry sticks for fuel and an alpenstock, while Van Trump's load was increased by a change of underwear, a boxed mirror and eighty feet of three-eighths-inch rope.

The route chosen led them onto the ridge now known as Success Cleaver at a point two miles from camp. On the left lay South Tahoma Glacier, which was so seamed and broken with crevasses that they decided to go much higher before attempting to cross it to the comparatively smooth ice of Tahoma Glacier. At an elevation of 11,000 feet, they descended onto the ice in order to pass around the rocky cliff that separates the two glaciers, and night found them sheltering on a wet ledge, protected from rock-fall but not from the cold. A little fire made with the sticks carried up from below enabled them to have a hot meal, after which they tried to sleep (Riley's deerhound kept their feet warm, but could do nothing toward keeping the cold, penetrating wind from their shivering bodies).

The climb was resumed at 5:30 on the following morning, after a cup of hot coffee had eased the chill. At six o'clock they were at last upon the ice of Tahoma Glacier, though more than 2,000 feet below the summit.[16] The morning was spent plodding upward on the medial line of the glacier, a route which, though appearing smooth, became increasingly more difficult. The icy inclines steepened as they advanced, making much tedious step-cutting necessary, and deep crevasses lying across the line of travel became more numerous, deviating them more often. As a result, noon found the climbers below the 13,500 foot elevation,. with the worst traveling ahead.

Hot coffee was again managed by building a fire in a hole cut in the ice to protect it from the wind. Whether or not the deerhound shared their lunch of meat, bread, and cheese, we do not know, but he almost provided himself with a tidbit nearby. A small, dark object was seen running over the upheaved ice toward the climbers, and was soon identified as a mouse. The dog went out to examine the little animal, which jumped into his mouth while he was sniffing at it from a foot away. Bowser was so surprised he dropped his inadvertent captive and lost him into a crevice. All were at a loss to explain the presence of the rodent in such an inhospitable and unlikely environment.

After lunch, the arduous climb was continued with renewed energy. Lower on the glacier the soggy, new snow would bear their weight if they stepped carefully, but the colder air at higher elevations resulted in dry snow through which they floundered with increasing difficulty. By 5:00 p.m., the steepest climbing was behind them, though progress was slow through two feet of lightly crusted snow which had already cut the stitching in Van Trump's boots and let in the finely powdered snow—his feet were very cold.

The North Peak was close upon the left, but could not be approached directly because of a large crevasse which forced the climbers to plod eastward to the saddle connecting it with the main summit before turning toward their goal. Another hour gained them only 200 feet of elevation, with at least that much left to achieve, and so, after a brief consultation, they turned their backs at sunset upon the bleak, windswept peak that would certainly cost them "hardship and suffering, such as men should not voluntarily incur" were they to conquer it. Instead, they made their way through gale-driven snow to a steamy refuge on Crater Peak, where they arrived at nine o'clock— exhausted. But cold, fatigue, and disappointment quickly receded into sleep.

Morning brought a burning thirst which caused Dr. Riley to set about melting some ice in the coffee pot. While so employed he found the sheet of lead left on the summit

126

by George Bayley in 1883. The neatly engraved names of Bayley, Van Trump, and Longmire remained legible, despite the thick layer of corrosion, and another name, "W. Fobes," had been scratched on the back.[17] The find stimulated Van Trump to make a search for the copper plate left by himself and Stevens in 1870, but again it was not located.

Van Trump and Drewry went to the west rim of the small crater, placing the mirror there by wiring it to a rock facing in the direction of Yelm and Olympia. They returned, shivering with cold, to find that Dr. Riley had made some observations on the physiological effects of altitude: pulse up from seventy to 110 per minute, and respiration at forty times per minute, which led him to estimate that the lungs required eight times more air at that elevation than at sea level.[18]

The dangers and difficulties of the route by which they had ascended, together with the desire of Riley and Drewry to see Paradise Park, decided the climbers in favor of a descent by way of Gibraltar. After leaving a small board with their names on it among the rocks on the bare eastern rim of the small crater, the descent was started. Van Trump led off across the large crater and had taken only a few steps until he was floundering in deep snow. Abandoning the short cut, they made their way around the exposed rim, passing the place where the Enumclaw party had raised their flagstaff, of which nothing remained except the butt end "broken off even with the top of the pile of stones and almost as smoothly as if it had been sawed off."

Deep snow gave them an easy descent to Gibraltar, where they each crossed by a different ledge. The remainder of the journey to Camp of the Clouds was increasingly painful for Drewry and Van Trump, both of whom had the inflamed eyes of coming snow blindness while the latter was suffering also from frostbitten feet. Luckily, they found encamped there a party of would-be mountain climbers from Centralia who "hospitably made them welcome to everything eatable and drinkable in their possession."[19]

The Centralians noted that the poor dog was as badly used up as his companions and they wisely concluded their own equipment was inadequate for the ascent of a mountain which could so batter man and beast. Though Van Trump and Drewry were snow-blind the next morning, the party made the difficult fifteen-mile trip across moraines and snowfields to their base camp, where they arrived after dark. Both unfortunates were laid up in camp for several days; in fact, Van Trump was yet *hors de combat* on August 17 when Riley and Drewry met three adventurers from Orting while exploring the Puyallup River drainage.

The worn and defeated mountaineers reached Yelm on the 23rd, and the two Olympia men returned to their city "under cover of darkness that night." It was a fitting entry for Dr. Riley—entirely in keeping with the lack of character he proceeded to display.

Two days after his return to Olympia, Dr. Riley gave the Tacoma *Ledger* correspondent some details of the ascent. Now there should have been honor enough for Riley merely in being a participant in a bold ascent, which, though it failed narrowly in its objective of reaching the virgin North Peak, yet was a gallant attempt in the face of terrible conditions. Failing of that larger view, his ego should have been content with the fact that he certainly came off physically best man. But he was not one to gracefully accept less than he had proposed to have, and consequently readily explained away defeat on the North Peak as the fault of his companion, Van Trump.

As the doctor told it:

> There were 900 feet yet to do . . . Here Mr. Van Trump became sick and exhausted and the last ration of whisky and beef tea was issued . . . Darkness hung over them, the wind blew at the rate of sixty-five miles an hour, with the thermometer 25 degrees above zero . . . . Mr. Van Trump at this time lost the power of speech. He could not be kept awake and was compelled to walk in his sleep. His circulation had almost ceased and he became black and purple. All the inducements to urge him to the crater . . . were used, and the point was reached . . . .[20]

That tactless release was followed by a less derogatory one in the Olympia *Tribune*,[21] and a nearer-to-the-truth account in the Tacoma *Ledger* which, incidentally, carried the humorous sub-title, "After wading through snow sixty feet deep they soak their heads for a few days!"[22]

The veteran climber, Van Trump, might have let Riley's slanderous, publicity-seeking stories go unchallenged if he had not been personally maligned. He countered brilliantly with a signed article in the Tacoma *Ledger,* giving a well-written account of the ascent and concluding, in part: "when an event or occurrence somewhat out of the ordinary course cannot be made interesting clothed in the plain and simple garb of truth, so that imagination, hyperbole and exaggeration must be called in to aid, the event had better be left untold."[23] In the public opinion, Van Trump won his point and Riley's reputation began a rapid decline.[24]

At that point, the former climbing companions can be left until the next season.

It is a curious coincidence that the western aspect of Mount Rainier, which had not even been seriously considered by climbers prior to 1891, was ascended twice in that year. Remember the three young men from Orting who met Riley and Drewry on the upper reaches of the Puyallup River? Frank Taggert, Grant and Frank Lowe were returning from an unsuccessful attempt to reach the summit, and they were persistent enough to try again and make it. Their ascent is virtually unknown and the account of it here was written for the Orting *Oracle* by Frank Lowe in September, 1891:

> We left Orting about three weeks ago and went by the way of Lake Kapowsin and Baker's store[25] out the old Succotash Valley trail, till we struck the Rainier fork of the Nisqually. From there we took what is known as Indian Henry's trail following it for about five miles, then left it and cut our trail towards the mountain. We were 3 days reaching the great glacier on the northwest side of the mountain from which flows the Puyallup River.[26] Here we made our permanent camp for the trip and, after arranging the camp, started out goat hunting, but failed to get any.
>
> We left camp next day and started up the mountain to an altitude of 12,000 feet, far above the clouds, which we could see floating majestically below us, and left our guns. After leaving our guns, of course we saw many goats. On the side of the mountain towards the North Peak at about four o'clock we came upon a flock of goats, 5 in number. They allowed us to

129

come very near them, and I thought I would catch one of the little ones. I picked up a rock and threw it at one of the old ones. The old fellow got mad and commenced pawing the ground and shaking his head, but finally turned and walked off. We threw more rocks at him, but he quickly turned the tables on us and we had to run for the big rocks where we were kept prisoners until they went down the hill. Again we started upward over the last hill between us and the North Peak, only to encounter another flock of 32 goats. They did not run and we did not crowd them. We started back to camp at half past five, but, after reaching the place where we left our guns, we had great difficulty in finding our way owing to the very thick cloud which enveloped us. We followed our tracks, however, and reached camp at 8 p.m.

The next morning we started out to find a horse trail down the Puyallup River. We saw one deer and several whistling marmots on our way back, but met with no incident of interest.

On the morning of the 17th of August we packed our things preparatory to the ascent to the top and started upward. We traveled over icy fields until 10 o'clock when we came to a large crevice which it was impossible to cross, and as the "white cap" was hanging over the top of the mountain we concluded to give up the attempt that day and return to camp.

On our return we met Dr. Riley and A. Dreury, of Olympia, who told us they had just returned from the top of the mountain. They also told us that P. B. Van Trump, of Olympia, was in their party, but was then laid up in camp with his feet badly frozen. The Doctor's face had been frost-bitten and the skin had peeled off, leaving his face almost raw. They told us that if we were going to the top, to take plenty of blankets, for we would need them. That night (the 17th) it rained all night and a heavy snow fell on the mountain so we deferred the ascent until the 19th.

On the morning of the 19th the weather was clear. So we packed up and left camp at 6 o'clock. We traveled over ice until noon when we stopped for dinner and, continuing our journey, traveled over a ledge of rock until 3 p.m. when we again found a snowy stretch in front. At 4 o'clock we stopped for a breathing spell and figured out we would reach the summit by six; but we kept on until eight o'clock, and we were still far below the top. We resolved to camp there for the night and, finding a sheltering rock to protect us from the winds and the rocks which are constantly rolling down the side of the mountain, we dug out the snow to secure a place to lay down. We had carried a bundle of dry wood for fires and igniting a few sticks put some ice on to melt. It was slow work and was a long time before our coffee would boil. Never did a fire seem better, for it was piercing cold. At 9 o'clock it grew still colder and we were almost afraid to go to sleep, so we wrapped our double blankets around us and sat up until 2 a.m., silent and cold. We felt too miserably cold to talk much, but at that hour Frank and Grant lay down and tried to sleep the rest of the night. I sat up till morning and when the sun came out it seemed to us like a hot stove, its feeble rays were so grateful.

After breakfasting we started for the summit at 7 o'clock. The snow was frozen hard, compelling us to chop our steps until 10 o'clock when we reached the distance of a hundred yards, but distances are very deceiving at this height, and we found it was fully a quarter of a mile. After hard climbing we finally reached the top of Mt. Tacoma, and, discovering smoke or steam issuing from a part of the crater, we made our way hastily to that point and were soon enjoying the genial heat.

After warming ourselves we began to look around. There are three peaks; from the top of the west peak to the east one[27] there is a distance of about a mile and a half. The slope on the west peak is an easy one and no difficulty in walking is encountered; but the east peak has a snowy slope of 45 degrees. We discovered a piece of sheet lead or lead plate about $3\frac{1}{2}$ x 8 inches, with the names of P. V. Van Trump, G. B. Bayley and James Longmire scratched thereon with the year 1883, and more recently on the 14th of August, 1891, the names of P. V. Van Trump, Dr. Riley and Alfred Dreury. There were no signs of any others having been on the summit and we added our names by means of a knife on August 20, 1891.

As I said, there are no signs of any other persons ever having been on top of Mt. Tacoma and I should think that if parties have ascended to the summit as they claim to have done they would leave some signs or articles to show the fact. It is not a trip one can make before breakfast or in a day, and a name or record on that mountain is worth having. From the smaller crater, where we found the lead, it is about 50 to 75 feet lower than the highest point. On attempting to cross the snow we broke through and several times I had to raise myself up by my hands and staff, the wind blew like a cyclone and it was difficult to stand up against it. It took us 46 minutes to walk around the rim of the large crater on the bare rocks and we judged it was about a quarter of a mile in diameter. It is full of snow up to 75 feet of the rim and on the northwest side the heat keeps the snow melted, leaving the rocks bare.

The small snow peak seen to the right from Orting is about 40 feet higher than any other point and is as round on top as a ball, being almost 75 feet in diameter at the top.[28] It appears to be solid ice and snow. We were on the summit for about three hours and traversed all parts of the bare rocks we could find, but found no signs of the Brown party having been there or the piece of flag staff that is said to have been found by the

131

Van Trump party. The only evidence we found of others having been on top of the mountain are those I have mentioned.

We started on our downward journey at one o'clock by way of the west peak. We were five hours going down sliding on the snow and ice, till we reached the Puyallup glacier. From there on we went over the great cracks in the ice by means of ice bridges and made our way rapidly to the camp, arriving there safe and sound. Our dog greeted us as though surprised at our return and our horses seemed pleased to see us for they had been alone for thirty-five hours.

The next day we lay in camp all day keeping wet cloths over our eyes. It seemed as though they would burst every moment and we were all in considerable pain. Frank Taggart was in the worst fix for his face was almost raw from a run on the ice. We started home on the 22nd of August, and were three days getting to Orting. We were greeted with many warm handshakes and many questions about the mountain; but it would take a book to describe all the many incidents and sights experienced in climbing old Mt. Tacoma to the summit. We have many rare relics, consisting of rocks of many kinds and colors and other curiosities which we shall long treasure as mementoes of our trip to the top of Mt. Tacoma.[29]

The west side route was thus proven feasible for novices as well as experts, with the former faring somewhat the better by their direct approach. Yet, the rough treatment received by both parties on the Tahoma Glacier leads one to believe Frank Alling may have had the better idea when he proposed to avoid the rock-fall danger on Gibraltar by cutting a new trail across it, instead of looking for an alternative route.[30]

The 1891 climbing season was concluded with an attempted ascent by members of the Oregon Alpine Club. Charles Gove returned with five members of that organization for an ascent by the Gibraltar route. They began their climb from Camp of the Clouds at 12:40 a.m. on August 23rd, with the intention of climbing directly to the summit without using a high camp.[31] It was a good idea and would later be successfully employed by hardy climbers, but it didn't work in 1891. William Steel was too soft for the pace, giving out below Gibraltar, with the result that the party did not reach the ledge in time to pass it safely.[32]

Gove's party was easily defeated and went away, probably unaware that their failure was symbolic of the organization they represented. The Oregon Alpine Club was even then dying: regular meetings had ceased and a schism over objectives had divided the membership beyond hope of reconciliation. In due time the Portland *Oregonian* would provide a brief obituary,[33] thereby closing the record of the first mountaineering organization in the West.

During 1891, the Public Land Survey was pushed up the valley of Nisqually River through completion of work on two townships,[34] thereby leaving a graphic record of Succotash Valley as it was that fall. On the plat returned by the surveyors is a notation, "Ashford's Field" where a town would later be, and there is a road shown running down the valley from Kernahan's ranch for four miles, for some of those lonely settlers were cutting a way toward civilization. There were others building roads too, as the following will testify:

> Mr. Fletcher, who lives at Brown's junction on the Nisqually river, about 12 miles from the Mineral creek mines, says that the wagon road is now open from Tacoma to Brown's junction . . . . There is a hotel at Brown's junction. A road will be opened this season from Succotash Valley to the junction, which will be the route from Tacoma to the mountain. All the people in this region do their trading at Tacoma, and with an improvement in the roads will continue to come this way, though Centralia is making great efforts to take the traffic to that city.[35]

Thus passed the year 1891—a year memorable not because of the opening of a new route to the summit (for it was a dud like Oscar Brown's publicity stunt and Dr. Stafford's volcano), but for the viable germ of a great idea that was planted in the public mind. Let us not forget that it was at Yelm in August of that year when an earnest little man who loved the mountain as few ever have wrote these words:

> One subject in connection with the mountain and its surroundings which has often occupied the writer's thoughts, and one in which many more of the citizens of our state ought to be concerned, he would introduce here, with the hope that it may call the attention of others to it and excite their interest and influence — and that is, a movement looking to the setting apart of the mountain and its extensive and beautiful parks as a state institution or reservation, or, failing that, as a national park.[36]

And the by-line he wrote under was P. B. Van Trump!

About this time the simmering controversy over the name of the mountain came to a violent boil. The old mountaineer, Van Trump, raised his voice in moderation, suggesting that, if the historic name of Mount Rainier was doomed to banishment, it should be replaced with a true Indian word rather than Theodore Winthrop's invention, Tacoma. Regretfully, he offered "Tah-ho-mah" as a substitute, but he was not heard any more than King Canute.[37] There would be much bitterness engendered and many dollars would be spent in the attempt to graft the name of a city upon the mountain its populace had laid claim to.

James Longmire had located at his "Springs" in 1884, yet he did not claim the land until October 5, 1889, at which time he filed a notice at Vancouver, Washington, indicating his intention to enter a placer mining claim in Pierce County. His claim was designated Lot No. 37 and surveyed as containing 18.02 acres of land, and in due course, he received a patent which made him truly the owner of "Longmire's Springs" on February 6, 1892.[38]

The climbing season of 1892 begins with the announcement that Frank Lowe and Frank Taggart (Orting men who climbed the previous year) would climb to the summit in order to set off twenty pounds of red-fire at midnight on the Fourth of July.[39] However, their schedule went awry by more than a month and they were not the first up.

Perhaps it was the backwardness of the summer that delayed the Orting party, for at mid-month, Mr. Julius Schweigart reported Camp of the Clouds as "the only green place in all Paradise Valley, which is as yet entirely covered with snow."[40] Other prospective climbers are identified in the newspapers, including the rivals, Dr. Riley and P. B. Van Trump, but the season's first ascent went to a dark horse party.

The leader, George L. Dickson, gained his experience the hard way, for he had already made three attempts to reach Mount Rainier's summit. He first tried his luck in 1889 with the Dodge party, which shed exhausted climbers all the way to the top of Gibraltar, where the last three were blocked by a crevasse they were not prepared to cross.

In 1890, he did no better from contributing his pack and coat to Nisqually Glacier while crossing the "ledge," and in 1891 the total exhaustion of a companion on the summit dome was all that kept him from the goal.[41]

Dickson's 1892 party consisted of his brother William H. Dickson, Holland W. Baker, and Dr. W. G. Cassels of Tacoma, and W. E. Daniels of Sumner. From Tacoma they took "the cars" to Spanaway, going from there by "the stage" to Baker's store on the Little Mashel, where they arrived the same night.[42] The next morning they loaded their equipment on two packhorses hired from Baker and followed the rough road which Pierce County had cut through the forest to the Nisqually (seven and one half miles) the previous summer. Reaching the river at noon, they had dinner at Sully's before turning onto the well-beaten Yelm trail, which they followed to Kernahan's ranch for a night camp.

Longmire's Springs were reached at noon on the third day, and, after bathing and drinking copiously of the various waters, they pushed on to the lower edge of Paradise Park. Dickson was disappointed to find that a forest fire had destroyed much of the alpine forest the previous fall, quickly attributing the destruction to the deliberate vandalism of "some miscreant." Actually, the fire was accidental.[43]

By camping early at the edge of the ravaged forest, the party had a chance to rest and dry footgear which had become soggy crossing melting snowbanks. The following afternoon a move was made to another site three miles farther up the ridge at the highest point of timber line, where they established a base camp they later christened "Camp Success."

An ascent of the mountain was begun on the morning of July 28, with each climber equipped in the manner of that day with alpenstock, caulked shoes, and a thirty-pound pack containing provisions and blankets. In addition, a hatchet, a coil of half-inch rope and two canteens (one filled with whisky which proved unpalatable on the heights) were taken. The hot sun softened the snow and made traveling hard work. Noon found the party near Camp Muir, and by sunset they were close under Gibraltar. There they huddled together on the narrow ridge, wrapped in wool blankets that failed to blunt the cold wind blowing up from both sides at once. It was a shivering, sleepless night which left them "all very glad when it became light enough to travel. By unanimous consent this place was named 'Camp Misery'." Thus was the poetic Camp of the Stars banished by a more realistic name which is yet quite appropriate.

The ascent was resumed at 5:30 a.m., with the top of Gibraltar reached at eight o'clock and the crater rim at 10:15 a.m. The climbers remained on the summit until noon on July 29, 1892, and, while making a circuit of the larger crater, found a stone cairn which contained a tin box with the names of a Seattle party which had reached the summit in 1889.[44] The members of the Dickson party added their names to it before descending. The debris with which they were bombarded while passing under Gibraltar provided several narrow escapes but they reached Camp Success without mishap at five o'clock that evening.

The Dickson party reached Tacoma several days later, with this prophetic comment: "Someday in the near future, when a good road has been made into Paradise Valley, it will become one of the greatest summer resorts on the coast." In later years, as a Pierce County road commissioner, George Dickson helped turn his own vision into reality.

Unfortunately, Dr. Riley reached the mountain before Van Trump. On the 20th of July he passed through Yelm in a light wagon, accompanied by George Jones. Baker's store was as far as they could travel in that manner; beyond it, they rode horseback — one on a draught animal and the other on a saddle horse which had been led behind the wagon. The third animal served as a packhorse.[45] Riley probably retraced the route by which Van Trump had guided him to the west side of the mountain the previous year, for he and Jones were soon encamped in the alpine park at the end of Indian Henry's trail. The two Olympians left their camp at about the same time the Dickson party left Camp Success, but their more difficult route up the Tahoma Glacier threw Riley and Jones behind, with the result that they reached the summit six hours after the Tacoma men had left it. The night was spent in the small

136

crater, where they may have had company — for the first time there were two parties on the summit at once.

The others were Frank Taggart and Frank Lowe, the Orting boys who had intended to celebrate the Fourth of July on the mountain top. An article in the Tacoma *Ledger* hints that they accompanied Riley and Jones, but I think it is more likely that Taggart and Lowe were following their route of the previous year and that the two parties met where Riley and Jones swung onto Tahoma Glacier.[46] Anyhow, they joined forces for the assault on the North Peak. Leaving the shelter of the small crater at eight o'clock the following morning, the four climbers easily crossed over to the coveted peak on a crusted snow, arriving there at 10:30 a.m., July 31, 1892.[47]

The stay on the North Peak was a short one as a temperature of 27° F., combined with a violent wind, made its summit untenable. Placing a tin box containing their names under a cairn of stones, they made a brief, and very inaccurate, geological investigation before retreating.[48]

The North Peak was conquered. After twenty-two years of striving, the feat was done with comparative ease — a mere walk on crusted snow from the central peak! And what of the hero? His reception in his home town must have been a disappointment to a man so hungry for acclaim.

The Tacoma newspapers gave him more space than those of Olympia but neither were particularly enthusiastic. As before, Riley's story was a curious mixture of pseudo-science and thrilling adventure. Science took a rebuff in his "barometric estimate" of the height of Mount Rainier, whereby he was able to "put it at near 15,000 feet," thus showing a fine disregard for the well-known fallibility of the best barometric work. The basis of his interest in the mountain's height lay in its up-and-down fortunes. Mount Rainier's presumed first place among United States mountains, on the basis of the 1870 determination of 14,444 feet, was soon challenged by adherents of Mount Whitney (which they considered to be 14,898 feet high). With an elevation of 15,000 feet, Mt. Rainier would again be in the lead.

Nor did the Spirit of Adventure fare better when Dr. Riley recounted an accident which befell him. High on the mountain, he tumbled fifty to seventy-five feet when a ledge of rock crumbled beneath him — and had the presence of mind to photograph his descent with a Kodak which was strapped to his belt. The editor of the Tacoma *Ledger* considered Riley "thoughtful" in documenting his adventure.[49]

With that we can forget a mountain climber whose accomplishments could not gloss over the defects of his character.

As for Riley's companions, the two Orting boys, Frank Taggart and Frank Lowe, received a chastisement in the public press when it became known that they carried away as a souvenir, the lead plate left in the small crater by Van Trump, Bayley and Longmire in 1883.[50] The influence of bad company, perhaps?

Dr. Riley's carelessness with the truth leads to a suspicion that he may have had something to do with the wild stories which were circulated that summer concerning the route to Mount Rainier; yet, in all fairness to him, there is no evidence he was their author. The burden of those tales was the dangers which must be met on the way to the mountain — wild animals, raging floods, snowstorms and rough men — such were the perils pictured. The invidious publicity was finally scotched by an anonymous "tourist," who set the record straight.[51]

Despite the rumored dangers, many interesting people went to Paradise Park during August of 1892. Among them was a Tacoma violinist by the name of Professor Olof Bull, who favored the excursionists encamped there with some charming moonlight musicales. He also had fond hopes of reaching the summit with his violin, but failed to get beyond Gibraltar ledge that season, due to the lateness of the day. On the descent, a young lady of the party was injured by rolling rocks; her account of that trip to the mountain is a charming narrative, deserving something better than obscurity.[52]

However, a member of what I will call the Griggs-Hewitt party did reach the summit. W. W. Seymour made his successful ascent under the guidance of Henry Carter, a man employed by James Longmire to build a trail up

Paradise River.[53] Carter guided on the mountain for a number of years, and it is said that he always climbed without caulks in his boots. In 1893 he homesteaded on Bear Prairie, where prominent drainage ditches remain to mark the site of the "90 acres of good hay land" he once had there.

There were others who came to vacation, but not to climb: among them a Mrs. Moore, who was an artist; our old friend, E. S. Ingraham, with a party of five from Seattle; and Judge Calkins, who accompanied the Griggs-Hewitt party of ten from Tacoma.

Among Mount Rainier visitors was a man who came neither to climb nor to vacation, but only to take the quickest possible look. He was Professor Garrett P. Serviss, a noted lecturer and scientific editor of the New York *Sun*, who had a commission from the Urania Club of New York to present a series of lectures in eastern cities on the scenic wonders of the Pacific Coast. General Passenger Agent Fee, of the Northern Pacific Railroad, induced him to include Mount Rainier on his itinerary. He arrived in Tacoma July 9th, was taken on an all-expense-paid trip to the northwest slope of the mountain, arriving at Crater Lake on the 10th, and was back in Tacoma in time to make the evening train on the 12th, having "inhaled and absorbed . . . sufficient inspiration to give him a brilliant conception of the beauty and majesty of the wonderful monument."[54] He had also absorbed sufficient Mount Tacoma propaganda from his guide, Fred Plummer, to embarrass him greatly at a later date.

During the summer of 1892, the Longmires built a new trail up Paradise River to the valley and parks at its head. The work was done by Len Longmire and Henry Carter, who was familiarly known as "Step-toe." One day Elcaine Longmire and his daughter, Maude, walked up from the springs to see how the work was going, meeting Carter near a waterfall. After the men had talked a while, Carter turned to Maude and said, "Miss Modesty, how about naming these falls for me?" Maude told him she would think it over and let him know later.

She stayed at the falls while her father and Carter walked

on up the trail, and when they came back she said, "I've named your falls — they are Carter's Falls." He protested, "That's not fair!" but Elcaine told him he would have to put up with it, since he had asked that the falls be named for *him*. Carter laughed, then took his axe, blazed a tree, and wrote the name on it.[55]

Near where the trail entered Paradise Valley, an old campsite was found, and a piece of rusty iron stuck into the trunk of a nearby tree identified it as the place where the injured Van Trump was left in 1870 while Stevens went for help. Len pulled the old ice axe down, later replacing its rotted handle with a new one.[56] He had what was yet a strange piece of climbing equipment at Mount Rainier.

A short item in the August 6th issue of the Tacoma *Every Sunday* announced that Dr. Grant S. Hicks, Dr. George B. Hayes, Dr. Van Marter, of Rome, Italy, and Julius Schweigart planned to leave for Mount Rainier the following day with the intention of ascending it. The issue of the 13th notes that they left on schedule, but without Dr. Hicks, thus setting in motion what can best be called the "doctoral ascent" (remember, there was an editorial ascent in 1890).

Dr. Hayes had developed an enthusiasm for mountaineering through contacts with some of the earlier climbers, and Dr. Van Marter had considerable experience in the Alps (where he had made two ascents of the Matterhorn and one of Mont Blanc); Schweigart was to be the guide. The trip to the mountain was uneventful and the three men lounged about their camp in Paradise Park for several days, enjoying the scenery while acclimating themselves (among their conditioning climbs, was the first recorded ascent of Pinnacle Peak). The day they selected for the summit climb began with perfect weather, but clouds began to form around them as they climbed. By evening they had reached the foot of Gibraltar and the wind was blowing a gale with no visibility except through occasional rifts.

Encamping themselves on a narrow ledge called the "Eagle Nest," they crawled into rubber-covered sleeping bags to spend a cramped sleepless night buttoned up, heads and all, against the worsening storm. With the coming of

140

daylight, they retreated to Paradise Park, their lives saved by those first sleeping bags used on Mount Rainier.[57]

Several days were spent recovering from that frigid ordeal, then they were ready for another try, but without Schweigart, for he was through with mountain climbing! The second time, the two doctors had better luck. Leaving camp at ten o'clock in the morning, they reached the old high camp at Eagle Nest by dark. A 4:45 a.m. start was made from there on a clear but windy morning; Gibraltar was passed by seven o'clock, and the top was reached at 10:00 a.m., August 20, 1892. Dr. Van Marter was the first foreigner to climb Mount Rainier and he appears to have been a competent mountaineer.

Julius Schweigart did no more guiding, but there were others. We note that "Mr. J. G. Semple . . . accompanied by the guide, Henry M. Sarvent, climbed to the top, but did not reach the highest point," and also the following advertisement: "G. C. Fletcher . . . who accompanied the Griggs party to Paradise valley, is prepared to accompany parties to the mountain, being familiar with the route and the needful preparations for the trip."[58]

Soon after the 1891 attempt to reach the North Peak, Van Trump received a letter from his old friend George Bayley of San Francisco, asking that he be counted in on any future tries. Early in the summer of 1892 they corresponded, arranging to make an ascent during August.[59]

And so, on August 15th Bayley and his daughter arrived at Yelm, and the next day the two mountaineers departed for Mount Rainier, while Miss Bayley remained a guest at the Van Trump home. Well mounted and with a pack-horse to carry their outfit, they traveled rapidly through country which was changing so much with settlement that "a faithful description of it today would differ greatly from a description of it last year."[60] They found that trails had given way to roads along much of the route, with other portions under contract. Ten and one-half miles of county road were being constructed between Sullyville, nine miles beyond Eatonville, and Kernahan's "Palisade Ranch."[61] Gangs of men were busy "corduroying miry places, building bridges over streams, felling trees, and digging or blasting

141

out stumps." While he approved of the appearance of stores, hotels, schools, and farms along the road, Van Trump regretted that the saloon, "inevitable, though undesirable" had arrived before the church.

Indian Henry's trail was again followed up the Rainier Fork into the beautiful park which now bears his name, and there they pitched the white wall-tent at sundown on August 18, on the "extreme edge of the meadow nearest the snow line."[62]

The ascent of the mountain was begun on the 20th. One horse was picketed near camp, one was turned loose to graze, and the third was loaded with the packs to serve as transport as far as he was able. But Oneonta was not a mountain horse; a mile from camp he had to be unpacked and turned loose to rejoin the others. Shouldering their heavy packs, each containing a blanket, extra clothing, and food for two days, and distributing a spirit lamp, a large sheet-tin reflector, a hand axe, and rope between them, they took up their alpenstocks to scramble on up the rocky ridge half a mile before descending to South Tahoma Glacier on their left.

Crossing the glacier so much lower than previously proved no advantage, for its surface was seamed with crevasses, and they arrived at their night camp on a ledge of rock at an elevation of 11,000 feet no earlier than by the previous route.

After a "cold and comfortless night," the climb was resumed up Tahoma Glacier, but on a course to the right of the 1891 route. It was a serious mistake which entangled them in a hopeless labyrinth of crevasses wasting both energy and time. They finally returned to the medial line of the glacier, and with better traveling, reached the foot of the central peak by evening.

The calm, bitterly cold air soon froze the damp shoes of the two climbers stiff on their cold feet. Following the precepts of that day, they took off their shoes, rubbed already cold feet with snow, and later suffered frostbite as a result. The remainder of the plodding climb to the small crater was made by moonlight, and their objective was reached between ten and eleven that night.

Van Trump and Bayley were soon lying on the warm inner wall of the crater in the mouth of the steam cave which had sheltered them in 1883; but this time there was no violent gale overhead to buffet them with cold drafts and colder dashes of flying snow. It was calm on top the mountain.

Sunrise turned the summit of Mount Rainier and its companion peaks into "domes of burnished gold," and the mountaineers were soon up exploring in the continuing calm. Van Trump was perplexed at not being able to find the lead plate left there in 1883, for he had seen it in 1891, and was yet unaware that Taggart and Lowe had carried it off. The mirror was in place on the west rim of the crater, but it had a thick coating of ice upon its face which had to be carefully picked away before it could again cast the sun's rays toward Yelm and Olympia.

The trip to the North Peak was started at 7:00 a.m. Less than two hours' travel over firm snow was required to reach that coveted goal, but it was no longer the virgin peak it had been a short time before. On the rocks which form the base of the eighty-foot high snow mass or "Liberty Cap" which is the true summit, a piece of sheet lead was found in a crevice, and at another point a cairn of rocks had been erected. Bayley deposited their sheet of lead by it and Van Trump wired the bright tin reflector to the rocks so that it pointed westward. After a brief and disappointing view of the mist and smoke-shrouded landscape to the north and west, the two climbers began their descent at 11:00 a.m. on August 22nd.

Bayley was in the lead at the steepest point on the Tahoma Glacier, about 2,000 feet below the summit, when a most unusual accident occurred. He stepped on an icy spot where his caulks failed to hold, going down so suddenly that he lost his grip on his alpenstock, which remained sticking upright in the crusted snow. Tumbling and spinning on the icy slope, Bayley slid rapidly down the medial line of the glacier toward a great crevasse which bisected it 2,000 feet below. Part way down, he managed to get into a sitting position, with his feet forward, and then attempted to slow his speed by the friction of his gloved hands and by digging

143

his heel caulks into the ice. In front of the crevasse was a nearly level strip of snow and Van Trump, watching anxiously from above, had some hope his friend could stop himself there. But Bayley's momentum was too great; it carried him into the abyss which barred his path.

Shocked by what had happened, Van Trump took his friend's alpenstock and began a cautious descent toward the crevasse. He stopped to rest a hundred yards down the slope and was surprised to see what he thought was a small man (Bayley was small and wiry) standing on a level place two hundred yards below the point of disappearance. As he watched, the movement of arms and legs could be plainly seen as though the other were walking up to meet him. But the figure made no forward progress, and Van Trump finally realized that desperate hope and a distraught imagination had created a cruel mirage from the heat waves rising above a rock.

Van Trump reached the edge of the crevasse somewhat to the left of the point where Bayley went in, shouting down with no real hope of receiving a reply. To his surprise Bayley answered from directly below. The crevasse was twenty feet wide at the point where he had gone in, and his momentum had carried him across it in an arc which slammed him against the farther wall, after which he had dropped sixty feet to a narrow ledge below. He was fortunate in receiving no worse injury than some broken ribs, and was able to crawl along the ledge to the point where Van Trump found him.

Bayley was soon pulled out of the crevasse with a rope, but he was in such pain that he could travel only very slowly, with Van Trump carrying the packs and assisting him down the steep places with the rope. Despite the fortitude and pluck of the injured man, evening found them at the exposed ledge where they had encamped on the ascent. Shelterless, they suffered through a terrible night there.

The following day Van Trump began to succumb to snow blindness, his constant adversary on the mountain, and that, with the numerous crevasses to be crossed or bypassed, kept them from reaching the rocky ridge above their base camp until evening. Dusk and a storm caught them there, so they decided to camp rather than risk losing their way in the fading light and driving mists.

Van Trump improvised a shelter from a double blanket and the alpenstocks, then laid in a supply of wood, and there they remained through a windy night and the following forenoon. Abandoning hope of a clear-up, they decided to try to reach camp despite the poor visibility. In that they were lucky, finding camp late on the afternoon of the 24th after some anxious searching.

The picketed mare was in a pitiable condition. The rope had caught on a bush, snubbing the animal so closely that water and grass could not be reached, while horse flies had covered her with blood-crusted sores from their bites. The other horses had drifted off in the storm.

Oneonta was found the following day more than a mile from camp but could not be caught. It was necessary to abandon the camp equipment and start at once for help with Bayley mounted on the weakened saddle horse. Between her slowness and his inability to stand jolting, it was late at night when the unfortunate mountaineers reached the Palisade Ranch. The hospitable Kernahans took the battered climbers in and soon had Bayley's broken ribs set. During the eight days he and Van Trump rested there, a party returned to their camp, recovering the abandoned horses and equipment.

The trip to Yelm on horseback caused Bayley's ribs to pull apart, despite the tight binding, so that he was suffering greatly on arrival. The next train took him to Portland where he received adequate medical attention.

In the article written for the Tacoma *Ledger* immediately after his return from the mountain, Van Trump spoke of his "recent and final ascent of the old mountain."[63] With the statement that he was content to rest upon his laurels (which were many in the public estimation), he gave up mountain climbing. And what better end could there be for twenty-two years of climbing than to be able to say, as he could, "I am pleased to note that at last there seems to be a public move in the direction of petitioning the government to set aside Mount Tacoma and its environs as a national park or reserve.[64] I have for a long time advocated this idea, and my feeble pen has done what it could in the matter . . . ."

As Van Trump ended twenty-two years of mountain climbing, the ice axe he carried to the summit in 1870 went to the top for the second time.[65]

Lawrence Corbett stayed at the Springs with Len Longmire that summer. His sister Edith was one of the three women who had reached the summit, so Lawrence kept urging Len to take him to the top. One day Len surprised him by saying, "Let's go up now," and they did, reaching the top on September 3rd, which was Lawrence's birthday.

They camped in the crater and that night there was a heavy fall of snow which covered up some of their equipment, including Van Trump's old ice axe. The boys had a difficult time getting off the mountain. Len never did find the lost ice axe, though he searched for it many times, and it probably is there yet, buried deeply under the consolidated snows of many seasons—a splendid immortality!

Yes, the mountain fever was certainly epidemic in 1891 and 1892, and out of the confused events of those two wonderful years came very great results. Judge Calkins returned from his Mount Rainier outing in a thoughtful mood. He had talked with Van Trump, and with Bayley, while the latter was recuperating from his injuries at Kernahan's ranch; thus, he brought home some shrewd observations obviously influenced by the views of those ardent conservationists. News of the Oregon effort to have Mount Hood set aside as a forest reserve was a sufficient cue for the judge. In an interview with a reporter from the *Ledger,* he made this proposal:

> The government ought to set aside about twenty sections of this land, including Paradise valley, and make a national park of it. It is unsurpassed for beauty and attractiveness and would make a very great summer resort. Mount Tacoma looks grander and more magnificent the nearer you approach it, and is said by mountain climbers to be the giant of the world for climbing purposes.

The reporter appropriately titled the article, "Save Paradise Valley,"[66] which struck a note of real public interest. It

146

was followed by an editorial optimistically promoting the national park idea.

There is considerable talk at the present time in favor of a National Park in Pierce County including our grand old Mount Tacoma, and many of the neighboring volcanic peaks. Scenic experts are loud in their praise of the wonders of this region, and the many hardy tourists who have ventured to the snow slopes have returned enthusiastic admirers of this wonderland of the northwest. It is proposed to petition for a reservation about twenty-five miles in width and thirty miles long which would include all the region of perpetual snow and the beautiful, grassy slopes for which the Mountain is famed. It will include the headwaters of the White, Carbon, Mowich, Puyallup, Kautz, Nisqually, Paradise, and Cowlitz Rivers together with twelve enormous glaciers which head these streams. It will also contain more than a dozen beautiful lakes and twice as many waterfalls and cascades.

In the southern part of the proposed park is a magnificent range of volcanic peaks known as the Iatoisch [sic] range, of which Eagle peak is a prominent member. To the northward of Mount Tacoma is a range of peaks similar in characteristic called the Saluskin range.[67]

It is doubtful if another region could be found upon the globe where such a wealth of magnificent scenery is included in such a small area and so near to civilization. Its field for the naturalists is unlimited and the scanty supply of paintings and photographs which have been made fall far short of doing justice to the majestic grandeur of the region and offer many inducements to artists. If the park is secured, roads, trails, and possibly railroads will soon be under way for the benefit of those who wish to avail themselves of such privileges, and fair accommodations at Paradise Valley may be expected next season.

Mr. Fred G. Plummer is the promoter of this enterprise, and he says he needs only the moral backing of the people of Pierce County to make it successful. That great explorer, Lieut. Frederick Schwatka, has taken interest in this matter and will probably visit in person the proposed Park and lend his valuable influence in our behalf. To such a worthy object there can be no opposition and everything points to a successful issue.[68]

Another item published the same day went into more detail on some points. It proposed a park comprising 500 square miles (twenty miles east-west and twenty-five miles north-south, and informed readers that the data gathered by Fred Plummer and Lieutenant Schwatka, "of Arctic fame," would be used to memorialize Congress for the establishment of a national park. The lieutenant was also to "write a book and magazine articles . . . illustrated with pictures of a high character so as to attract public attention to the mountain and its claims upon the traveling public."[69]

It soon developed that Lieutenant Schwatka was no simon pure altruist — he was only in the scheme for the money. The Tacoma Commercial Club had proposed to back the Mount Rainier exploration, but, as the good weather of fall dwindled away, the free-lance lecturer became more evasive and his price increased. Finally, it came out that he had "a proposition of a similar nature from another city under consideration."[70] After that, silence; Tacoma's plan to memorialize Congress for a national park at Mount Rainier was dead.

Again, the anticlimax: a Professor L. F. Henderson, "The well-known botanist for the Washington World's Fair Commission," had hoped to be a member of the Plummer-Schwatka expedition.[71] With the collapse of that venture, he decided to go on a one-man expedition to the alpine meadows of Mount Rainier. While encamped at the mountain, the equinoxial storm caught him, apparently unprepared. The result was that the professor lost his entire outfit and wandered for hours in a blinding snowstorm. Nearly dead with cold and hunger, he escaped by following down a stream which brought him at last to Longmire's Springs.[72]

The result of it all, the stunts and climbing, work and scheming, was an upsurge of interest in Mount Rainier that brought to light the "pearl of great price"— recognition of the national park caliber of the mountain and its surroundings.

CHAPTER VII

MEN OF THE MOUNTAIN

1893-1894

The proposal to establish a national park at Mount Rainier took hold with astonishing rapidity, particularly in the vigorous young city of Tacoma. There it soon became the focus of a popular movement which was violently impassioned, hydra-headed in its objectives and less effective than it might have been by reason of its narrowly commercial viewpoint. A mighty furor was raised, and soon subsided, leaving true accomplishment in calmer, more directly purposeful hands.

Mid-January found Fred Plummer's map of Mount Rainier completed and the Tacoma *Ledger* gave it a front page spread under the inaccurate title, "The First Map of the Summit of the Old Mountain Ever Published."[1] Three days later P. B. Van Trump wrote a letter to the editor, commenting that "on the whole Professor Plummer's map represents a pretty fair birdseye view of the mountain. There are doubtless errors in detail, but that is to be expected . . . ."[2] Engineer Plummer acknowledged Van Trump's well-meant remarks, inviting others familiar with the upper slopes of Mount Rainier to inform him of such errors they might have observed in his map. All would have been well had he not added the following:

Permit me to make a correction of the generally accepted narratives of the so-called first ascent. Messrs. Stevens, Van Trump and Longmire were not the first to make the ascent. General A. V. Kautz made the first ascent, although he has never claimed it, and only claimed 12,000 feet. I obtained from him a detailed account of his route, and do not hesitate to make this

149

statement, without fear of successful contradiction: On the 16th day of August, 1857, Lieut. August V. Kautz, Surgeon R. O. Craig and Private Dogue, all of Fort Steilacoom, Wash., reached the crest of Mount Tacoma, and stood upon the ridge connecting Peak Success with Crater Peak, and were the first white men to reach [the] summit.[3]

Plummer could not deliberately have barbed Van Trump any deeper than by that blunt assertion on a debatable point. However, he did not do so with the intention of stirring up the touchy old mountaineer, though it might appear so. Through coincidence and tactlessness an unfortunate argument was opened.

The coincidence arose when Major General August V. Kautz, retired since January 5, 1892, arrived in Tacoma the day after Plummer published his map. Europe, where he had been vacationing, must not have been to his liking, for he left his family to winter in Rome and returned to the Puget Sound country.[4] Imagine his surprise at finding that the subject of conversation was a map of Mount Rainier upon which his route of 1857 was very inaccurately shown. Kautz went to Plummer protesting that the map did him an injustice by terminating his ascent at the 12,000 foot elevation below Peak Success. He soon convinced the engineer that he had actually reached the saddle between that peak and the main summit, assigning it a height of 12,000 feet on the assumption that Mount Rainier's height was 12,330 feet, as determined by Lieutenant Wilkes in 1842;[5] and the statement Plummer made claiming Kautz was first upon the summit was merely his attempt to rectify what he considered an injustice.

It was tactless on Plummer's part, for he could have revised his map to correctly show Kautz's route without taking the matter to the newspapers, or he could have made a public statement of the facts, as he had them from Kautz, without injecting an opinion regarding the claim of Van Trump and Stevens to a "first" ascent.

As we know, Van Trump was definitely hypersensitive concerning the 1870 ascent; thus his reaction to the claim Plummer had made in Kautz's behalf was immediate and vigorous. He wrote the editor, "If the *Ledger* will open its columns to a limited argument on this question it will prove of interest to many and settle once for all an important chronological event in the history of the mountain." The editor was agreeable, so Van Trump opened with a long dissertation to the effect that the ascent of Kautz and his party was incomplete, since they did not reach the highest point, and that he and Stevens were entitled to the honor because they did. He cited a precedent in the case of the first ascent of the Matterhorn, where Tyndall's ascent to an elevation of 13,970 feet in 1862 — only 810 feet below the summit — did not dim the luster of Whymper's subsequent conquest of the peak. He ended with the challenge, "If it can be proved that General Kautz was 'the first white man' to stand on the lofty crest of our grand old peak, and to look down into its smouldering craters, I shall not regret it, and will be the first to say: 'Let honor be paid to whom honor is due'."[6]

Fred Plummer's reply on the following day was more an appeal to emotion than to reason. He began with a statement of his regret that Van Trump "shows some feeling, and attacks me rather than the facts in the case"— which certainly was not true — and he continued by badgering the definition of a "summit" to his purpose. His reasoning was that the first man to cross the Cascade Mountains reached the summit, though he was in a pass; therefore, that Kautz had reached the summit of Mount Rainier because he attained a saddle from which the mountain sloped in two directions. It was an entirely phony argument, which he ended with a fine disregard for the obstacle which the last four hundred feet of a great mountain can represent, by saying, "when one has climbed the slopes and turned the crest, he is on top, and is entitled to that honor, whether he splits hairs by standing on Crater Peak or not."[7]

The effect was to commit General Kautz to the defense of a claim he would never have made! To the reporter who interviewed him at the Fife Hotel, he said: "I think that I fairly ascended the mountain. While I admit that I was not upon any one of the peaks, I did stand on the divide between the middle peak and the south peak, between 6 and 7 o'clock in the evening and saw the sun going down between the Olympic range . . . If I had had time I could have gone to any of the peaks, I think, from the appearance of the ground."[8] His honest statement, made without rancor, fails to support the contention that he reached the summit; but it does provide substantial evidence as to where the *ultima thule* of the 1857 expedition was. Place it at an elevation of 13,950 feet, and that will rectify the error which grew out of Lieutenant Wilkes' faulty determination of the height of Mount Rainier in 1842.

Returning to the focus of the argument, Van Trump based his rebuttal on the dictionary and the accepted practice of mountain climbers with regard to claiming "first" ascents.[9] It was conclusive, for Plummer slipped out of his indefensible position by suggesting that Kautz take over from there,[10] but he, too, dropped the argument.

Fred Plummer's map embroiled him in another direction. Among those suggesting corrections was E. S. Ingraham, who wrote from Seattle to object to some of the place names used. He did not approve of the changes which eliminated "Edmunds" (for Senator Edmunds of Vermont), "Blaine" (Secretary of State under President Harrison), "Tyler" (first manager of the Tacoma Hotel), and "Carbon," "White River" and "Inter" as names of glaciers; nor did he care to see "Lace Falls" again Spray Falls, the name Bailey Willis had given it. Ingraham did suggest that no more names be given to the features at Mount Rainier until such unhonored heroes as Stevens and Van Trump received their due, and in that he was seconded by the editor.[11]

The suggestion that his name be given to one of the glaciers of Mount Rainier brought an immediate remonstrance from Van Trump, who said: "Now, while I feel flattered and honored by the suggestion, I must earnestly and sincerely protest . . . I am opposed to applying the names of men, even great ones, to mountains, rivers, glaciers, or any of the sublime things in nature."[12] It was his feeling that only the names of the rivers they form, or authentic Indian names of the area, should be used to denominate the glaciers; otherwise, let them be "nameless here forever more!"

Bailey Willis was another who disapproved of Fred

152

Plummer's name changes, and he took prompt and effective action by referring the matter to the United States Board of Geographic Names. As a result, the Board approved the place names used on the Willis-Berghaus map of 1886, which thus became the official map of Mount Rainier.

Another of the comments brought out by Plummer's map is interesting because it fills in our knowledge of events southeast of the mountain. Late in June of 1889 a Chehalis party led by J. T. Forrest reached the valley of the Williwakas (they used that name) by way of Cowlitz Valley. Plagued by bad weather they were able to get no higher than 8,000 feet on the mountain, but they did find the Cowlitz box canyon, where the waters passed "through a gap ten feet wide and 200 deep."[13] Forrest returned to the Williwakas with another party in July, 1892, and again failed in an attempt on the mountain. They were stopped at an elevation of 11,000 feet on Ingraham Glacier.

It was at this time, while the Plummer-Van Trump controversy raged, that the agitation to establish a park at Mount Rainier bore the first fruit. As we have seen, Van Trump suggested the idea in August of 1891, and Judge Calkins called for action a year later. The resulting popular movement ballooned with astounding rapidity, coming to the attention of the Federal Government within five months. And so, it probably was with no really great surprise that the people of Tacoma opened their newspapers on January 26, 1893, to read that "Secretary Noble has granted the necessary order for the reservation for all time of several thousand acres in this country, including that most symmetrical, if not the grandest mountain peak in the world—Mount Tacoma."[14]

On another page of the same issue, some of the hopes of the park's promoters were indicated. They were confident the new park would be twenty miles wide and twenty-five miles long; that an annual appropriation of $30,000 each from the national and state governments, with "something from the county," could be secured for its management; that a good carriage road could be built to Camp of the Clouds for $15,000; and that a good mule trail, ninety miles in length, could be built around the mountain.[15] On the

153

following day, the *Ledger* enthusiastically assured its readers the "reservation proclamation probably will be issued today," and then went on to delineate boundaries as follows: beginning at the SE corner of Section 32, T19N, R7E, then running true south thirty miles, true east twenty-four miles, true north thirty miles and true west 23.79 miles[16]—the park's area had thus been increased from 500 to 720 square miles overnight!

Those announcements were premature, for an impasse had been reached — what should the name of the new area be? The existence of two factions, representing Tacoma and Seattle, had been recognized for many years; the Tacoma group being in favor of calling the mountain by the new, but supposedly aboriginal name of "Tacoma," while the Seattle group was for retaining the old name of "Rainier." Adherents of the two groups had already passed the point of politeness, let alone compromise, and were stridently insisting the reservation should bear *their* favored name. The fight was immediately taken into the offices of the Department of Interior, where action to set aside the area was delayed while a solution to the name difficulty was sought.[17]

A way out of the dilemma was finally found by dropping both the contested names and substituting one unassociated with the quarrel or its factions. On February 12, 1893, the area embracing Mount Rainier was designated "The Pacific Forest Reserve," and its status as part of the reserved lands of the United States was confirmed by a proclamation of President Harrison, February 20, 1893,[18] just two weeks before the expiration of his term of office. The disappointment of the Tacoma group was great, but they accepted the name in a spirit of resignation — drawing some joy from the fact that it was not called Rainier.[19]

Nor were the actual boundaries of the new area as satisfactory as those proposed earlier. The western boundary was drawn five miles nearer to the summit of Mount Rainier in an effort to exclude some mining claims vaguely known to be located near the headwaters of Mashel River. It was a shift which eliminated a three-mile strip of the present national park from the reserved lands, leaving some doubt as to whether or not the lower reaches of the west side glaciers lay within the area. On the other side of the mountain, the eastern boundary was established more than three townships beyond the summit of the Cascade Range, and there was a southward extension of one township.[20] Thus Mount Rainier stood barely within a great tract of wild country roughly thirty-six by thirty-nine miles in extent, half of which lay in eastern Washington and formed no part of the mountain's environs.

Those poorly established boundaries would later be a source of difficulty, but Tacoma had its park — for so the citizens of the city considered the Pacific Forest Reserve; yet it would take a few years and some doing to accomplish something more than mere reservation. It needed to be "enabled."

Had the readers of the *Ledger* given serious thought to a remark Kautz made to a reporter who interviewed him, they would have had some inkling of the tremendous obstacle to be surmounted before their favorite area could truly become a park — protected, financed, and managed. He said:

> I suppose if the park is reserved the government will survey it and agree upon some plan of taking care of it. I don't know whether it will care to go to any further expense in parks or not, it has so many of them now. You know the Yosemite valley has been made a national park, and now if they make Mount Tacoma a park somebody will want the base of Mount Hood set apart, and some one else will want Mount St. Helens, and so on, and in a little time the government will have more parks than it can handle well.[21]

His logic was a common one in that day and formidable, too!

In such a climate of heightened interest, should we be surprised to find the mountain fever taking an organizational turn? A mountain climbers' club of sorts may have been formed in August of 1892 under the name Tacoma Alpine Club,[22] and there was much talk that fall concerning the possibility of again forming an organization similar to the Oregon Alpine Club, but nothing came of it then. It was James Van Marter, Mount Rainier's first climber with European experience, who revived the idea. In a long letter

155

to the editor of the *Ledger,* he made an eloquent plea for a broad organization, outlining in it the pleasures of mountaineering, the benefits to be derived by individuals and Tacoma as a whole from a thorough exploration of the state's mountainous areas, development of a competent guide service, preparation of adequate˙ maps and guide books, and by furthering road construction and the establishment of hostels.[23]

Van Marter's challenge, "Let us organize an Alpine club in Tacoma, and do it now," brought the immediate response he desired. Four days later the following call was issued:

> Tacoma, Feb. 9— We, the undersigned, having before this date climbed to the summit of a prominent peak in either the Cascade or Olympic mountain ranges, in Washington, do hereby call a meeting of all persons who have reached the summit of any peak of either range to meet at the office of Fred G. Plummer, at No. 917 C. Street, room No. 12, on Saturday afternoon, February 11, 1893, at 3 o'clock to form a Washington Alpine Club.
>
> James Wickersham, Henry M. Sarvent, William W. Seymour, George B. Hayes, Fay Fuller, Fred G. Plummer, Albert Whyte, John G. Sample, Deborah S. Wickersham, A. [J.] Schweigart, H. S. Griggs.[24]

The initial meeting resulted in formation of a temporary organization under the chairmanship of Judge Wickersham, and plans for a meeting at the office of Dr. G. B. Hayes, in the Tacoma Theatre building February 25 at 8 p.m., for the purpose of formal organization. Among those present, in addition to the sponsors, are three familiar names — General August V. Kautz, Professor Olof Bull, and Dr. Van Marter.[25]

With perfection of the organization but two weeks away, trouble developed. It took the form of gossip to the effect that Fred Plummer was candidate for the position of commissioner of the newly created Pacific Forest Reserve, hinting that he was influencing members of the temporary organization to elect him president of the Washington Alpine Club to further his chances for the federal plum. More damage was done by a rumor that women were to be excluded from the organization. Neither charge was true, if subsequent denials were sincerely made, yet the effect was to alienate a number of mountain climbers.

While the important charter meeting was being held on the 25th—at Plummer's office again—some dissenters were attending a meeting called by Dr. Van Marter at The Tacoma. Those who attended the Washington Alpine Club meeting approved the articles of incorporation filed that afternoon, adopted a constitution and by-laws, and elected officers: General A. V. Kautz for president; Fred Plummer for secretary; and C. B. Talbot for treasurer — while the dissenters concluded their meeting by appointing a committee of seven to approach the Tacoma Alpine Club with the proposition that they reorganize and admit new members.[26] The effect of schism was to rob the new Washington Alpine Club of strength it could ill afford to lose, particularly Fay Fuller's support.

That enterprising young lady was a feminist, and a most unusual one. After seven years of school teaching, the last four at the Longfellow School in Tacoma, she turned seriously to newspaper work, becoming Tacoma's first woman reporter — covering the waterfront, the courts, and the markets, walking miles from one end of town to the other holding her skirts out of the dust and mud.[27] As if that wasn't boldness enough, she also took on the job of port collector, reputedly the first woman to hold such a position in the United States.[28] Meanwhile, Miss Fuller drilled with the Women's Guard, a group of twenty young women who had formed a team to partake of the healthful benefits of marching and rifle calisthenics!

Fay Fuller went East in April to cover the Chicago World's Fair for her father's newspaper, and the "Mount Tahoma" stories, and the miscellany of her "Mountain Murmurs," disappeared from the *Tacomian*. Perhaps that was a greater loss than the fact that she remained aloof from the Washington Alpine Club.

The damage was offset for a time by close association with the Tacoma Academy of Science, with which joint meetings were held. General Kautz, P. B. Van Trump, and James Longmire were featured speakers, and the early meetings were well attended. Late in June another meeting was held

157

at Plummer's office which probably marks the zenith of the club. It was rife with optimism and ambitious plans, yet there was something lacking, for the Washington Alpine Club faded away to a quiet end two years later.[29] Time has sunk that once promising organization into oblivion, for it is in no way related to the Seattle club which later adopted the name and now wears it successfully, quite unaware of its ill-starred predecessors.[30]

Plummer's map and the controversy it inadvertently aroused made good newspaper copy in 1893, as did the establishment of the Pacific Forest Reserve, and the formation of the second Washington Alpine Club; yet if we may judge from the space given it, there was another topic of equal interest to those stricken with the mountain fever — the completion of a wagon road to Mount Rainier.

For the genesis of the idea, insofar as there is a record, we go back to a statement in the Tacoma *Ledger* which credits "Engineer Plummer, who is an enthusiastic promoter of the project," with suggesting construction of a carriage road to Camp of the Clouds.[31] But that suggestion struck no fire; nor did S. B. Pettengill's later proposal that a road be built to enable "a line of buckboard stages to start from Yelm . . . and run in one day to the Paradise hotel," which he thought could be constructed of logs at slight expense.[32]

Successful ideas are like crop seeds; they have to be planted in good ground, and it was James Longmire, the proprietor of the Soda Springs, who found fertile ground wherein to plant the idea of building "the mountain road." Picture the tall, gaunt, old pioneer in his black broadcloth best at a joint meeting of the Washington Alpine Club and the Tacoma Academy of Science, asking their aid in securing a road from Kernahan's ranch to Paradise Park, "so that a buggy might get up there." His presentation of the tourist potential of Mount Rainier and Tacoma's advantageous position in regard to routes and distances, was sufficiently convincing to gain the following resolution:

WHEREAS, It is expected that between 50,000 and 100,000 tourists will visit Tacoma this year, and it is believed that if proper arrangements were made a large number of them would visit the glaciers, mountain meadows, and even the snowy summit of Mount Tacoma; and, whereas, a county road is now completed to within nine miles of Longmire's springs, on the Nisqually, and dedicated to the springs: now it is

Resolved, By the Tacoma Academy of Science and the Washington Alpine Club jointly, that it is to the best interest of Pierce county that the road be at once completed to Longmires springs, and the county commissioners are respectfully requested to so complete it.

Resolved further, That a copy of this resolution be sent by the secretary to the board of county commissioners to the chamber of commerce and to the commercial club, and that they be and they are hereby requested to lend their aid and assistance to the better development of Mount Tacoma park before the summer visitors reach Puget Sound.[33]

Before their enthusiasm could cool, the citizens of Tacoma were rudely jolted by the news that the King County commissioners were sending out surveyors to find a route from Seattle to Mount Rainier.[34] It probably was no better than rumor, and may have been only shrewd propaganda, but the immediate effect was to bring forth comments which were commitments to the cause. Secretary Denton spoke for the Chamber of Commerce when he said, "We want to be known the world over as a park city . . . why should we not profit by this — one of our great natural resources?" Chairman Fawcett, of the Board of County Commissioners, admitted "Mr. Longmire has asked us for $500 to be used in extending the road . . . I am in favor of allowing it."

Two days later the Chamber of Commerce urged immediate extension of the county road from Kernahan's ranch to Longmire's Springs,[35] and a joint meeting of the Chamber of Commerce, the Commercial Club, and the Academy of Science was soon held, resulting in the designation of a committee to approach the county commissioners with a plan. It was a proposal to expend $7,500 on construction of the road, with the expense shared equally by James Longmire, the county, and the citizens of Tacoma (through public subscription).[36]

But there was a flaw in the plan; or rather it was all flaw! The Pierce County commissioners had no funds which they could divert to such a project, the people of Tacoma proved less generous than they were enthusiastic, and Longmire apparently considered his share too great.[37]

Thus, nothing came of the idea for a month, except a plea from Van Trump for a really good road to the mountain, so as not to discourage tourists with a road "like the darkey's road to Jordan, 'ha'd to trabble,"[38] to which a certain Major Grattan replied to the effect that he didn't think people who were physically able to make the trip minded a "little rough rocking part of the way."[39]

Late in May the organizations interested in the road project gave up hope of help from the county commissioners, and it is obvious James Longmire shared in that estimate of the situation, for he already had a force of his own men at work turning the nine miles of trail between Kernahan's and the springs into a road.[40] There was a lingering interest on the part of the Tacoma groups — some talk of building the road as a toll road if subscriptions failed (which they did) — but it was Longmire who did the building.[41]

The work was completed in less than two months by the old pioneer with his crew of ten men (including Henry Carter and some of Indian Henry's people), a cook, and one wagon, at a cost of $100 per mile. The Roy correspondent of the *Ledger* thought he could well have spent more on the work, remarking "that a road which would suit him, with a small outlay of capital, would not command the admiration and approval of a modern tourist."[42] Though many would prefer to follow afoot as their wagons and carriages bumped over the low-cut stumps and teetered on the sidling grades, the road was a reality and its effect was very great.

The road has led back to Mount Rainier — what of events there? The winter of 1892-93 was a hard one with more snow than the oldest residents of the Puget Sound country could remember. As a consequence, spring was backward and summer much the same; the first parties to reach Paradise found the valley and the parks covered with snow.

For the first time, visitors were able to use the new Paradise River trail built by "Steptoe" Carter and Len Longmire the previous summer.[43] It provided a better crossing over Nisqually River, avoiding the dreaded "switchbacks" that led from the river to Canyon Rim, and so was popular despite the toll fee of fifty cents.

The Washington Alpine Club had plans for a Mount Rainier outing that summer of 1893, but all that came of it was an exploration of the northern flank of Mount Rainier by three small groups of club members.

Henry M. Sarvent and C. B. Talbot, both young Tacoma engineers, took to the field on August 4. While their route is unknown, it is likely they followed the old Bailey Willis trail to Spray Park, burdened as they were with a view camera which could be used to make 8x10 or 11x14 negatives, and with fifty glass plates of each size. Evidently their effort was a success, for the Tacoma *Ledger* reported their return on the 23rd with "some magnificent views of the mountain."[44]

The following day, Sarvent went back to the mountain accompanied by W. I. Lowry. It was a rapid trip which took them past "Crater Lake scooped out by glacial action" and on to the ridge between the Carbon and Winthrop glaciers. From there, they attempted an ascent of Mount Rainier by way of what is now called Curtis Ridge, only to be stopped by an abrupt precipice at an elevation they estimated as 12,000 feet. Thus they were quite as successful as the better-equipped and more experienced climbers who have followed them to defeat on that treacherous ridge. On the return to Wilkeson, they noted three moraines left by the recession of Carbon Glacier, and the name "Philo Fall" was given to a waterfall 250 to 300 feet high which they found below the moraines.[45]

Sarvent and Lowry were back in Tacoma on September 6, three days after a third party of club members left for Mount Rainier "to climb to the summit by the Wilkeson glacier." It appears that Talbot, too, had drummed up interest in an ascent of the mountain from the north side. The party he brought back consisted of C. P. Ferry, known as "the Duke of Tacoma," H. Shepard, Oscar Nuhn, John Garvey, and "Billy" Driver, a little Welshman who had been the companion of Bailey Willis in the early eighties, but, if anything was accomplished, it has escaped the record.[46]

161

Such expeditions could hardly be called club outings, and we are left to speculate on whether it was the discouraging weather or the growing disunity within the Washington Alpine Club which kept its members at home; whichever it was matters not at all, for we know what such an outing was like from an interesting diary kept by Mrs. Arthur F. Knight while on a trip to the mountain that summer. There were ten persons in the party—all members of the Narada branch of the Theosophical Society of Tacoma, and typical tourists of that day. Here is Mr. Knight's record of the trip as he later reconstructed it from his wife's diary:[47]

We left Tacoma August 11, 1893, Friday. Had a pair of horses and a democrat wagon. Mr. Schwagrel [Tacoma Park Commissiner] had a horse and buggy which was only large enugh to hold two so the rest of us rode in the democrat. We spent the first night at the Traveler's Home where we got a nice pan of milk and all slept on the hay mow.

August 12: Got up about five and reached Eatonville about 9 A.M. We had to repair the whiffle-tree of the democrat so we rested the rest of the day.

August 13: Reached Elbe about noon, had lunch, arrived at Ashford's at 5 P.M.

August 14: Reached Longmire's and put up camp near one of the Springs. We all took a bath in the Spring and felt quite rested.

August 15: Got up rather late and made a trip to the glacier up the Nesqualie River bed, the glacier looked like a dirty snow bank until you got close then you saw the clear ice with a cave underneath. We drank water from the dripping ice. We tried to climb up the glacier but the rocks began to roll so we gave that up. Got back to the Springs and after supper all took a bath. The water came to our necks and we were so light in the water that it took very little to support us.

August 16: Stayed at the Springs all day and the women did some washing. Had a hand made washboard. In the evening all went over to Mr. Longmire's and sang songs and had a fine time.

August 17: Three of us took a trip up the mountain at the side [Eagle Peak] and when we reached the top we had a wonderful view of the Mountain and could also see our Camp. The mosquitoes were so bad we had to return to camp.

August 18: Started after breakfast for the Coal Creek. Had a hard climb but were well repaid for the Falls were wonderful.

Went up high enough to strike snow, found some beautiful flowers growing close to the snow.

August 19: A disagreeable day, wet and foggy, but we decided to get to Paradise. Put packs on the horses and forded the river which was rather high, but finally got across, took the switchback trail, reached Frog Heaven where we found some pretty flowers. The party was pretty tired when we reached Paradise Park but it surely was a wonderful place, full of clumps of trees and flowers everywhere. The packs were too heavy for the horses so we had left three of the bundles down the trail, and it was too late to go after them, all we had for supper was hardtack and coffee without milk or sugar.

Sunday August 20: Went down and got the packs and came back with Charlie Arnold and two friends of his. Mr. Arnold was once a close neighbor of ours but had taken up a ranch somewhere above Ashford's and we were much surprised to see him. When we got back with the packs we had a feast. In the evening we had a Theosophical meeting. Had a good nights rest with a nice, level place to lay on and plenty of blankets.

August 21: Got up at 4:30 had breakfast and started for Gibralter. The snow was nice and hard, which made it easy to walk on. It was wonderful to see the Glacier hundreds of feet thick and on the other hand the Mountain side covered with beautiful flowers. We all got up to about 10,000 feet when most of them gave out on account of short breath. Myself and two others went up to Gibralter and then came back and all returned to camp through a dense fog.

Tuesday, August 22nd: A rainy disagreeable day so stayed around camp all day.

August 23: A lovely day so we all started for a tramp. Went first to Paradise Falls [Sluiskin Falls], sat down and rested, then walked along the ridge, saw the most beautiful flowers, red, white, blue and yellow, all blended together and lovely trees. We kept walking seeing wonderful new views constantly. Pretty little lakes with several outlets. Reached camp about 4 o'clock, found another party had arrived and pitched a tent a short way from us. We sang songs in the evening and made so much noise that three of them came over to join the fun.

August 24: This is a disagreeable day but I started out alone as no one would go with me. Went down the Paradise River quite a ways and discovered some beautiful falls. Came back and reported what I had found so after lunch the whole party except my Mother-in-law went with me down the Paradise River. We came to a roaring torrent which Schwagrel wanted to name Washington Cascades and as no one objected let it go at that. Below this we came to the Falls which I had previously found. These falls dropped about 200 feet and spread out fan shape, and we all went into ecstasies over them and decided

163

to name them Narada Falls.[47] George Sheffield and Ida Wright carved this name on a tree beside the falls. My wife and I and Jessie Barlow went lower down and came to another rapids which we named Tahoma Cascades. Then we all went back to camp. In the evening the party camping near us came over and we had a cornet solo and chorus singing.

August 25: Friday. Packed up and started back home about 9 A.M. Had an uneventful trip and reached the Springs about noon. We exchanged some of our provisions for a meal at Longmire's and we surely were hungry and it tasted good. Rested awhile then started, reaching Arnold's and took possession, some of us sleeping in the cabin and the others outside in front of a fire.

Saturday, August 26: Got ready to pack up when the horses decided they wanted some more Paradise grass so off they started and I could not catch them until they got to the gate. It was 11 o'clock before I returned with them. Reached the wagon we had left and hitched up. Most of us were now used to walking so we walked nearly all the way. We reached Kurneham's about five. Mrs. Barlow was acquainted so she went up to the house and soon came back with Mrs. Kurneham and daughter Ruth, who had some jelly and a big loaf of bread which we made short work of. They had a nice log cabin and they took us into the garden and we had a feast of currents and raspberries. That night we slept in the hay.

August 27, Sunday. Had a good trip all day. I drove most of the way and walked only over the worst hills, reached Eatonville and had a fine dinner at Carter's. Slept in the barn.

Monday, August 28. Started for home at 9 A.M. Had an uneventful trip and reached home about 10 P.M. none the worse for wear.[48]

Ten years had made a great difference; where Van Trump, Bayley, and Longmire forced their way through thickets and over fallen logs in 1883, unremitting effort had provided a wagon road, rough though it was, over which tourists could ride to the mountain.

As we have noticed, the summer of 1893 was not remarkable, except for its backwardness; and the effect of that was to create unfavorable climbing conditions on the upper slopes of Mount Rainier. A number of parties made unsuccessful attempts to reach the summit, among them a Tacoma group which included our friend of the previous year, Professor Olof Bull.

He again set out for the mountain in the expectation of

ascending it with his violin to give a concert upon the summit, but his luck was not with him. As a prelude to misfortune, the violinist fell into the Nisqually River while fetching water for the evening camp on the second day; then, soon after starting the following morning, his violin fell from the wagon, which ran over instrument and case. But, the smashing of his violin was not the end of Mr. Bull's troubles. He lost control while glissading a slope below Gibraltar, was dashed into some rocks and painfully injured, ending his mountain climbing for the season, with a ride on an improvised rescue toboggan.[49]

Professor Bull could draw some comfort from the fact that only one party managed to reach the summit that season. They were not experienced mountaineers, but only some surveyors afield "in the interest of a projected railroad" (probably the line which had earlier been proposed to tap the timber of the Mashel and Nisqually drainages).[50]

William Bosworth, of the firm of Ogden and Bosworth, was in charge of the survey crew, which left Tacoma accompanied by Arthur French, a "photographer and crayon artist" of that city.[51] The Tacoma *Ledger* noted that French contemplated a visit to Paradise Park, but it failed to mention that the others also had the mountain fever. No time was lost running line, for they were all at Paradise two days later.

And so, at 5:30 a.m. on the morning of August 25th, the four surveyors (Bosworth, Guy Evant, Walter Wolff, and Robert Shollenberger) left Paradise Park with French — all determined to reach the summit. They were well equipped by the standards of that day, but had no guide.[52]

By the time they had reached the Cowlitz Cleaver (which they knew by that name), the dust arising from the face of Gibraltar let them know the rocks were already crashing down. They did not relish a crossing of the "ledge" under such conditions and so made a high camp at two o'clock near the 11,000-foot elevation. Blankets were spread under rubber coats at the base of a rocky ledge, where they were "ready for the night, which was spent shivering and listening to the big masses of snow and rock falling thousands of feet at frequent intervals, causing a tremendous roar."

165

The climb was resumed early, but they worked themselves into a *cul de sac* below the ledge, where so much time was lost backtracking that the climb had to be called off. Returning to their base camp at Paradise, the climbers prepared for another try while nursing the sunburns they had received.

Conditions were again considered favorable on the 29th, so a start was made at 12:45 a.m., in the hope of effecting the passage of Gibraltar by the early morning light. Mr. Shollenberger had had enough and remained behind as the four climbers filed up the ridge by the light of a full moon. They fared better than on the first attempt and lost no time finding the route across the Gibraltar ledge. No particular difficulties were encountered until they were above the chute; there the surface of the ice resembled frozen billows and progress was slow, particularly for the photographer, French, who was laden with his bulky view camera.

The top of Gibraltar was not reached until noon and they rested there for an hour while lunching on canned beef washed down with draughts from a trickle of melt water. Refreshed by the halt, the climbers continued upward on a zigzag course through a maze of crevasses.

It was one of those almost windless afternoons which are so rare and so pleasant upon the summit dome and the climbers could see two tiny figures watching their progress from a point 500 feet above Camp Muir. P. B. Van Trump and F. J. Mosman, who had just completed a thorough exploration of Paradise River with G. F. Evans and C. A. Kernahan, had climbed to that vantage point to watch the laboring summit party.[53]

The climbers reached the crater rim at 4:35 on the afternoon of August 29, 1893, just fifteen hours and fifty minutes after leaving their Paradise camp. They immediately made themselves comfortable in the small crater, where they spent the night wrapped in rubber coats to protect their clothing from saturation by the steam.

An exploration of the summit on the following morning showed that Van Trump's mirror was yet firmly attached to the rocks on the west rim of the small crater, and Oscar Brown's flag was found, flagpole and all, at the edge of the snow on the north slope of the main peak. Despite the strong wind blowing then, the flagpole was carried back to its original emplacement and reset, after which French took a photograph of the flag pole and its accompanying pennant as it snapped in the gale. According to one account, the flag was not left there, but was carried off to be sold piecemeal on the streets of Seattle at a dollar for each fragment.[54]

Among the many relics of former ascents found in the craters were some tennis rackets which had been converted to snowshoes, leaving an interesting enigma — whose were they? By ten o'clock, the climbers had satisfied their curiosity, taken quite a few photographs, and were ready to descend the mountain. They had several narrow escapes while passing Gibraltar, and a near accident while crossing the head of Cowlitz Glacier, where a member of the party slipped, lost his alpenstock, and was only saved from a slide into a crevasse below by the rope he had hold of.

Darkness found the successful climbers safely back in camp at Paradise where Mr. Shollenberger was waiting with the kettle on the fire. So ended the only ascent of Mount Rainier made in 1893, a year when Paradise Valley had more snow on August 30 than it usually had in mid-July.[55]

Of course the controversy over the name of Mount Rainier continued unabated despite the 1890 decision of the United States Board of Geographic Names. Far from proving final, that verdict only drove Tacoma to a stubborn, and sometimes rancorous defense of its position. The "Rainier-omainiacs" were damned for supporting "that somebody or nobody called Rainier," and every shred of evidence or sympathy for Tacoma's cause was avidly seized upon.

Among those whose sympathy was enlisted, was Professor Garrett P. Serviss, who made that meteoric, all-expenses-paid trip to the mountain in July of 1892. In the presentation of his subsequent "Urania" lectures, he always referred to Mount Tacoma — and later found it necessary to defend his action.[56]

Another skirmish in the battle over the mountain's name took place at the Chicago World's Fair, of all the unlikely places! It appears that Tacoma was influential in the preparation of *The Washington State Souvenir,* a seventy-two-page booklet "authorized by the exposition to be distributed in the Washington State Building." Tacomans were pleased that their city "properly" received the most extensive write-up in the work,[57] which is not surprising, since publisher J. O. Hestwood was "willing to take the greater part of his pay for the issuing of the book in real estate in this city . . ."[58]

Tacoma definitely had the edge in published propaganda, but not in visual displays. Engineer Wood, of the Northern Pacific Railroad's mining department, prepared a fine map of the state to hang in the Washington building. On arrival, it was found that the mountain was designated "Mount Tacoma or Rainier"—a surprisingly broad viewpoint for a partisan. The assistant executive commissioner, Percy Rochester of Seattle, (who had bitterly and successfully opposed the name "Mount Tacoma") was reportedly furious. He employed a competent artist "to strike off the word 'Tacoma,' and told him to charge a good bill for his work and remarked that he would see he got his pay."[59] But then, Dr. Calhoun, who headed the Washington State Fair Commission, would not allow the map to be hung up (though it finally found a place in the back of the building as the result of a compromise). As for Mr. Rochester, he got the sack two weeks later as a more or less direct reward for his success in keeping the name "Mount Tacoma" out of the exhibit building.

While the controversy over the name of the mountain may not seem germane to this narrative, it really is, for it weakened the Washington Alpine Club through drawing the attention of its members from more proper and worth while objectives.

Of all the grand objectives of the Washington Alpine Club, only one was fulfilled. On August 9, Fred Plummer published a guidebook for Mount Rainier tourists.[60] It is the first such compilation and rather well done, also. The contents included a map of the route to the mountain, a table of mileages and a description of the stopovers along the road. Those pages give us an idea of what had been accomplished in the way of developing a tourist business centered on Mount Rainier.

An advertisement of the Lake Park and Tacoma Railway lists "Five Trains Daily to Lake Park where Tourists take Stages for . . . Mount Tacoma." The fare was twenty cents for the thirteen-mile ride from the end of the Jefferson Avenue street car line to Lake Park, beyond which lay twenty-one miles of rough, dusty, country road over Benson's Hill, past Tanwax, Clear Lake, and Ohop Lake to Eatonville. Though barely four years old, that town could boast a hotel "with accommodations for 60 persons," a general store run by T. C. Van Eaton, who offered "provisions, blankets, dry goods, hardware, boots and shoes, drugs, fishing tackle" and a "full outfit for camping parties." The tourist could also obtain the services of a blacksmith, barber, or cobbler there.

After crossing the wooden bridge over Mashel River, the road ascended Mashel Mountain to Meta, where a man named Baker had a post office and general store, then continued on an easier grade to the divide three miles beyond. The way led downward into "Succotash" Valley from there, reaching the Nisqually River at Elbe, forty-one miles from Tacoma. At that point there was another general store and two small hotels with a total overnight capacity of twenty persons, and Ashford's hotel, seven miles beyond, could accommodate another twenty.

If the overnight accommodations seem too few along the "Mountain Road,"[61] remember that they were used principally by those who traveled in the buckboard "stages"; tourists who had their own rigs usually camped by the wayside in good weather or sought the shelter of a friendly barn in bad.

The terminus of the road was at "Longmire's Medical Springs," sixty-one miles from Tacoma. The plain, almost comfortless accommodations and rough fare offered by James Longmire were quite acceptable after the hard trip in. As for the hot and cold "baths," they seem to have been thoroughly enjoyed by all who tried them.

Plummer's guidebook was undoubtedly a great help to the Mount Rainier tourists of that day, and it probably

169

had considerable effect in stimulating travel to the mountain.

Some of the other things done in the name of the Washington Alpine Club that summer were less helpful. Engineer Plummer undertook to re-measure Mount Rainier by angulation, reporting results between 14,900 and 15,100 feet for six observations of its height.[62] He thereby gave undeserved support to Dr. Warren Riley's contention that the elevation of the summit was 15,000 feet. Mount Rainier was again a contender for first place among the lofty peaks of the United States.

Nor was the exploration of the northwest corner of the Pacific Forest Reserve as fruitful as it might have been. While it resulted in a proposal to build a road from Wilkeson to Carbon Glacier along Carbon River, and a claim to the discovery of a new route to the summit (Lowery and Sarvent examined the Winthrop-Emmons route from Curtis Ridge, but did not try it), the road was neither needed nor practicable in that day, and the summit route was not new, for the three young men from Snohomish had climbed that way in 1884.[63] Between them all, the only accomplishment was to cause further confusion in regard to place names by their "additions to the nomenclature of the great mountain's surroundings."[64] The only park place names they can be thanked for are Echo Rock, Ptarmigan Ridge, Cataract Creek, and Moraine Park.

The publicity given Mount Rainier at the Chicago World's Fair sparked the interest of scientists in its glaciers. Some of Schweigart's photographs exhibited there aroused the curiosity of W. M. Davis, professor of physical geography at Harvard, and a favorable comparison of the mountain's glaciers with those of Switzerland, published in the magazine *Science,* brought a German geologist from Munich two months later.[65]

There was publicity of a different sort close at home. Under the title "Mt. Tacoma Vandals," the *Ledger* reported that "Seattle will make complaint to the Secretary of the Interior against the ruthless rustication of Tacomans in Paradise Valley."[66] In it lay the stuff of another acrimonious controversy—one in which Tacoma's citizens were far from united.

The complaint referred to stemmed from the observations of E. O. Schwagerl, one of the members of the Knight party which had recently returned from an outing in Paradise Park. Mr. Schwagerl was a "landscape engineer" whose genius had done much to beautify Tacoma's parks, yet without pleasing the city fathers, Thus, his trip to Mount Rainier was a sort of between-jobs vacation, after which he assumed the duties of superintendent of parks for the City of Seattle.

Mr. Schwagerl's professional training made him aware of the destruction then being wrought among the magnificent groves at Paradise Park. The careless hacking for camp wood and bed boughs, the appalling destruction of individual trees and clumps deliberately set alight to provide rocket-like conflagrations, and the hundreds of acres of desolated forest along the trail between the switchbacks and the park—all were goads to action. Schwagerl's aroused feelings soon found expression in a report to the Seattle park commissioners, and in it, he not only detailed the abuses but suggested the matter be brought to the attention of the Secretary of the Interior so that he might take steps to protect Mount Rainier's alpine wonderland.

The editor of the *Ledger* was quick to attribute the report to an "animus" resulting from Schwagerl's dismissal from the Tacoma park system. The idea that Tacomans were responsible for the alleged damage was scoffed at, since "no Tacoma persons, not even its vandals, would visit Mount Rainier even if they knew where to find it, which none of them do."[67] In the editor's opinion, the Tacoma Alpine Club could take care of the problem, if there was one.

The doubt was soon dispelled by Olympia and Tacoma parties. They confirmed the report that fire had done great damage at Paradise Park, but they shifted the blame to hunters, who made convenient scapegoats, as they were considered "vandals in all respects." Thereupon the *Ledger* reversed its opinion, admitting the government ought to take a hand in protecting the Pacific Forest Reserve — even to the extent of sending in soldiers![68]

171

At that point A. F. Knight rose up to defend Schwagerl. He was surprised that an "unworthy 'animus' should be charged against a late fellow citizen," reminding his neighbors that in the matter of their joint inheritance, it made no difference whether friend or foe reported the work of "sacrilegious ax and lurid torch." He thought that even if it was done through spite, they "should be thankful for the timely warning . . . even from a 'dismissed' landscape engineer."[69]

Desultory sniping over the vandalism issue continued in the Tacoma newspapers until mid-September, Fred Plummer, in his anxiety to prove that Tacomans could not be wholly to blame for the desecration of Paradise Park, provided an interesting summary of visitors during the summer of 1893 (probably taken from a register kept by the Longmires at their hotel). A total of 174 visitors were listed as follows:[70]

Tacoma	75	Orting	7
Puyallup	6	Mineral City	1
Seattle	8	Fern Hill	5
Olympia	5	Pierce Co.	9
Yelm	19	Elbe	2
Sherlock	1	Pennsylvania	1
Sumner	2	France	1
Alderton	2	Germany	1
Hillhurst	1	Unknown	19
Roy	9		

Van Trump's contribution to the argument showed his usual foresight; he recommended that Paradise Park be protected by posting warning notices, and by employing guards and men to make weather observations.[71] His recognition of the importance of fire weather was thirty years early.

For well over a year, the City of Tacoma had been attempting to obtain the water of Mashel River as a substitute for its inadequate and rather unsatisfactory supply — Spanaway Lake. But hopes for improvement were abruptly ended when Fred Plummer claimed water rights on nearly the entire volume of Mashel River. His subsequent attempt to sell his "rights" to the city for two million dollars got

172

him no deal.[72] That attempted holdup of his fellow citizens probably had much to do with the demise of the Washington Alpine Club.

While those "men-of-the-mountain" living almost in the shadow of the great peak pursued their various interests in noisy confusion — sometimes in harmony, but more often at cross purposes, sometimes with the public good in view, but quite often with only their own aggrandizement in mind; while they argued over the name of the reservation brought into being by their efforts, organized and fell apart in schism, promoted a road and failed to finance it, and dreamed great and largely fruitless dreams of "developing" their park; there were others at a distance who were accomplishing a grand design in an orderly and dignified manner.

That summer the Geological Society of America met at Madison, Wisconsin, and a committee was appointed on August 15 "for the purpose of memorializing the Congress in relation to the establishment of a national park in the State of Washington to include Mount Rainier, often called Mount Tacoma."[73] The members of the committee were Dr. David T. Day, S. F. Emmons, and Bailey Willis — two of whom we know very well!

The American Association for the advancement of Science, which was also meeting in Madison, followed suit six days later by appointing a similar committee consisting of Major J. W. Powell, the courageous explorer of the Grand Canyon, Prof. Joseph LeConte, Prof. I. C. Russell, B. E. Fernow, and Dr. C. H. Merriam — friends of conservation every one.

A third committee was formed for the same purpose at a meeting of the National Geographic Society in the City of Washington, on October 13. Its members were Gardiner G. Hubbard, Hon. Watson C. Squire (remember the name well), John W. Thompson, Miss Mary F. Waite, and Miss Eliza R. Scidmore.

Senator Squire of Washington State seems to have moved independently toward the goal, for on December 12, 1893, he introduced a bill into Congress under the title "To set apart certain lands now known as Pacific Forest Reserve,

173

as a public park, to be known as Washington National Park." It was supported by a petition, from the Seattle Chamber of Commerce and both were referred to the Committee on Public Lands.[74] Though S. 1250 was not reported back from committee (probably because the boundaries of the area it sought to establish coincided with the admittedly poor boundaries of the Pacific Forest Reserve), it marks the beginning of more than five years of sustained and finally successful legislative effort in Congress.

Senator Squire's bill should have been welcome news to Tacomans, but it was not so. The senator was considered a Seattle man, and, worse yet, he was identified with the "ditch gang," those promoters of the hated Lake Washington ship canal. Thus, his proposal received no mention in the Tacoma newspapers, though it was just what they had asked for previously.

As the year drew to a close, yet another committee was formed to further the establishment of a national park to include Mount Rainier. At a meeting of the Sierra Club in San Francisco, on December 30, President D. S. Jordan, John Muir, R. M. Johnson, George B. Bayley, and P. B. Van Trump were selected to represent that organization in the preparation of a memorial to Congress.

Represented on these dignified committees were men from the first, second, third, and sixth ascents of Mount Rainier, and some others who were not behind them in interest, and *they* were able to work together in complete unanimity.

A calendar year is an artificial boundary over which events march heedlessly. The action initiated by Senator Squire to create a Washington National Park was seconded by his colleague, Representative Doolittle, who introduced a similar bill on January 4, 1894.[75] His proposal was adversely reported by the Commissioner of the General Land Office in a letter of February 19, with the comment that the proposed bill was restrictive, reserving the area for the use of tourists alone, and that it failed to settle the claim of the Northern Pacific Railroad to granted lands lying within the Pacific Forest Reserve. It was the commissioner's opinion that legislation then before Congress would provide for

174

more adequate management of the reserved lands than could be obtained by giving the area park status. That appraisal received the endorsement of the Secretary of the Interior on March 6; as a result, H. R. 4989 also died in committee.[76]

On April 11, 1894, yet another of those quietly purposeful meetings to further the establishment of a national park at Mount Rainier was held. This time it was the Appalachian Mountain Club in Boston which appointed a committee of three members — John Ritche, Jr., Dr. Charles E. Fay, and Rev. E. C. Smith — with instructions to cooperate with the similar committees of the Geological Society of America, the American Association for the Advancement of Science, the National Geographic Society, and the Sierra Club in framing a memorial to Congress.

The cooperation of the three learned societies and the two climbing clubs was close and effective. The Geological Society of America seems to have taken the lead in preparation of the memorial to Congress, and Bailey Willis, who knew the mountain well, though he had yet to climb it, personally drafted the dignified text in which that petition was couched.[77]

Unfortunately, one of the men who had more than an ordinary share in the events which led so directly to the memorial did not live to see it presented to Congress. There was no storybook ending for George B. Bayley. That ardent conservationist, intrepid mountaineer and charter member of the Sierra Club should have retired from his successful business life to help preserve the wilderness he loved and promote the mountain climbing which was so irresistible to him, but that is not what happened.[78] While ascending alone in the elevator at his place of business, he put his head out of a side opening and was instantly killed.[79] So passed the man who brought John Muir's message north to Mount Rainier, there to root and grow!

The memorial to Congress was yet in the draft as another summer season opened at Mount Rainier. A winter of very heavy snowfall left the summer of 1894 nearly as backward as its predecessor, so that early July found Paradise Valley blanketed with snow "on an average ten feet deep in every

175

part of it."[80] And that was the wintery landscape which a large Seattle party, led by Major E. S. Ingraham, beheld upon arrival. However, Camp of the Clouds was free of snow; so a base camp was made among the early flowers which mantled that exposed ridge.

The party was the sort which delighted the congenial Ingraham, a mixed group of novices in whose company he stood out as a veteran mountaineer. The presence of an artist, a photographer, and a basket of carrier pigeons inevitably leads to the suspicion that publicity was an important objective of the expedition!

After three days spent at Camp of the Clouds awaiting an opportunity to make a summit climb, an early start was made on the morning of July 17th. Camp Muir was reached in good time, allowing the climbers to push on to Camp of the Stars. Ingraham described it as a "narrow ledge over-hanging a precipice," adding that "those not over five feet tall could lie at full length, but others of us had either the option of lying with our knees drawn up to our chins or shoving out our feet and allowing them to dangle in the wind,"[81] from which the Camp Misery of our day is readily recognized.

A night spent on that rocky perch is best described as a mingling of grandeur and bodily discomfort. Since it is sufficient to keep the overnight guest from sleeping in, sunrise found Ingraham's party pounding caulks into their shoe soles; then they were off up the rubble-strewn slope leading to the ledge, munching a cold breakfast as they climbed.

The narrow passage beneath the cliffs of Gibraltar, where "bold hearts faint and strong knees tremble," was soon negotiated with the aid of a hand line taken across by the guide, though equal caution was not exercised on the summit dome. It must have been a straggling ascent, for the first climber reached the crater rim shortly after noon and the last was not up until three o'clock. Major Ingraham was among the first, and he immediately skirted the rim to the snow mound which is the highest point. There he read his aneroid barometer (which had been carefully adjusted at the Seattle office of the Weather Bureau), determining the elevation of the summit as 15,500 feet at 2 p.m. on July 18, 1894.

Ingraham returned to where the party rested, awaiting stragglers, with the exciting news that Mount Rainier's summit was "the highest point in the United States, excepting Mount St. Elias."[82] That started an animated discussion concerning an appropriate name for the icy hummock between the craters. It was finally agreed that Mr. Hawkins' suggestion of "Columbia's Crest" was the most appropriate.[83] Thus, was the highest point of Mount Rainier given a name which has subsequently proven inaccurate, poetic though it may be.

The fact that Major Ingraham was inexperienced in the use of a barometer explains his readiness to accept a reading so out of line with all previous determinations of the height of Mount Rainier. He failed to recognize the effect of the developing weather condition (a storm was even then engulfing them) upon his instrument, and he failed to take into account the effect of temperature. To have had even reasonable validity, his observation should have been timed to correspond with a reading of a base instrument at a known elevation, which would have allowed an approximate correction of any general fluctuation in atmospheric pressure; also, the field instrument should have been carefully standardized to determine its performance characteristics under pressure and temperature conditions such as exist on high summits. Ingraham was asking a great deal of a notoriously variable instrument and the result was that he erred by 1,090 feet.

While the name Columbia Crest is inaccurate, if not inappropriate, it is well established by usage and official decision, so that regrets regarding it are probably wasted; nevertheless, it *is* regrettable that it supplanted a better name, Crater Peak, which was earlier applied to the central summit as a whole.

With the last of the climbers on top, essential formalities were performed. An American flag presented by E. Lobe and Company, of Seattle, and carried to the top by Leo Daft, was unfurled to take "possession of the crater in the name of Seattle," then the successful climbers posed while Mr.

177

Strickland photographed the largest group to ascend Mount Rainier to that time. There were fourteen in all: Major E. S. Ingraham, H. E. Holmes, Miss Helen Holmes, Miss Annie Hall, Miss M. Bernice Parke, F. W. Hawkins, Dr. L. M. Lessey, G. E. Wright, W. H. Wright, N. A. Carle, Roger Green, Jr., Ira Bronson, Leo Daft, and H. P. Strickland. With that group, three more women had conquered Mount Rainier.[84]

Remember the basket of carrier pigeons? One bird was released at Camp Muir, and after some difficulty, was finally started toward the home loft at Puyallup with a message. Another carrier and a "dummy," or common pigeon, were carried to the mountain top in order to send a second message from there. The dummy was released first, disappearing promptly, then a message was fastened to the homer and he was taken to Columbia Crest for release. However, the wind was so strong the bird did not want to leave the basket, but had to be driven from it, only to fly down into a small crater for shelter. Driven from there, he was chased from point to point on the crater rim for a time before disappearing in flight.

At about seven that evening, after the weary climbers had relaxed in their blankets on the warm earth inside the crater rim, both pigeons returned to perch forlornly on the rocks above. During the night it snowed, hailed, and blew in violent gusts, making the benumbed pigeons very easy to catch in the morning.

The party remained in the crater, imprisoned by gale winds, until the storm abated near noon; then they hastily abandoned the summit. As is often the case, the descent proved perilous, forcing them to use their "life line" for security. When safely back to Camp of the Clouds, the pigeons were again released, though the homer did not get away with his overdue message until the following morning.

As Major Ingraham's weary climbers were making their anxious descent from Mount Rainier's storm-buffeted summit, another party of climbers gathered on Mount Hood, a sister peak one hundred miles to the south, and there, on July 19, 1894, a new mountaineering organization was born. It was the "Mazamas,"[85] a group which took the place of the defunct Oregon Alpine Club, prospering where the earlier organization had not. Among the charter members was Fay Fuller.

With the ascent of the Ingraham party yet news, another group was making preparations for a climb. Mount Rainier's tourist possibilities had again stirred the interest of the Northern Pacific Railroad, with the result that Olin D. Wheeler, the editor of its publicity magazine, *Wonderland,* was authorized to make an official ascent.[86] The party which left Tacoma on July 31st was admirably organized — for its purpose. In addition to its editor-leader, there was Dr. Lyman D. Sperry, a lecturer from Bellevue, Ohio; a railroad photographer, George M. Weister, of Portland, Oregon; and the young Tacoma civil engineer, Henry M. Sarvent, whom we have already met. Ross Comstock, of Elbe, went along as "general assistant," which means that he was cook, camp boy, and packer.

A leisurely journey to the mountain found the party entering Paradise Park the afternoon of August 3rd. They were surprised to find snow blanketing the ground above the switchbacks, and Camp of the Clouds only "a little island rising out of the snow." They were also disturbed to find campers had already left their mark upon that beauty spot. Growing trees had been cut for fuel by many of those who preceded them, and we are made aware of their numbers by Wheeler's statement, "Indeed, at Camp of the Clouds, live timber is nearly all there is left for fires."[87] Yes, it is easy to agree with him that it *was* a perfect travesty to set aside such a magnificent domain without making the least effort to protect and maintain it.

Olin Wheeler was also concerned about a matter that appears to have troubled his successors very little — the perverse manner in which the public was applying a favorite name. He noted that Paradise Park was even then "called Paradise Valley . . . a most unfortunate name and entirely inappropriate," rightly adding, "It may, however, very suitably be applied to the small valley of the stream that flows from Sluiskin Falls." But never mind, only the United States Board of Geographic Names is that consistent.

While on the subject of vagabond names, even Wheeler

179

would have been surprised to know his Camp of the Clouds, on the low ridge extending southward from Alta Vista, was not the windy, inspiring campsite christened in 1886 and made famous by John Muir; but only the sheltered retreat of later and less hardy tourists. However, the new site could be as true to its name as the original, for Wheeler's party was soon cloud-bound, captives of the wreathy, drifting mist which "hid every semblance of mountain or valley."

On the morning of the third day, Sarvent decided to take a large pack up to the base of Gibraltar Rock where high camp was to be, and on his return that night, he reported climbing into bright, warm sunshine at 8,000 feet. The news was cheering and preparations were made for an immediate ascent of the mountain.

The climbing party of seven left Camp of the Clouds at ten o'clock the following morning, strung out behind Sarvent who was the guide by virtue of his previous experience on the upper slopes of Mount Rainier. Three thousand feet upward, the climbers halted by a trickle of snow water to eat lunch. While resting, they were treated to the grandest of all Alpine sights; the mists began to dissipate, opening fragmentary vistas of frowning crags, lonely snowfields, and chaotic ice falls magnified to awesome proportions by contrast with the vapid framing. Such a sight is never forgotten but looms forever in the memory.

Continuing on, the climbers followed outcroppings of volcanic rock and cinder where they could, plodding wearily through the sun-softened snow between them. On the great snowfield below Camp Muir the ascent became an endless succession of short traverses and short breaths which brought the party to the foot of the Cowlitz Cleaver as evening approached.

Not wishing to camp so low, the climb was continued to a little cove just beyond the Beehive. There a halt was made at seven o'clock by all except Sarvent, who stoutly climbed on to the foot of Gibraltar to recover the cache he had left there on the previous day. Wheeler says, "We all needed rest, and we all sat on the rocks for a time and took it." They were so tired even supper failed to interest them.

The starlit night was devoid of all familiar sounds; in

fact, of all sound except the muted rumble of an occasional avalanche, and that was no more conducive to sleep than the creeping cold. At 4:30 all were astir in the gray dawn, pulling on stiff shoes, exercising and preparing to ascend or descend, accordingly as the reflections of a comfortless night had dictated. From that high camp five continued, and two, who "felt they had lost no craters," returned.

An hour of careful travel over slopes glazed with ice brought the ascending party to the ledge, which was crossed with the aid of a fixed hand line placed there by Sarvent in 1893. The icy chute from the ledge to the top of Gibraltar Rock was safely climbed with moral support from "two-hundred feet of the very best window sash-cord" let down by Sarvent from an anchorage above, and noon found the climbers lunching at Camp Comfort. From there they went easily up the dome to reach the summit—Columbia Crest—at 2:15 p.m. on August 8, 1894.

The low-lying lands in every direction were hidden beneath a limitless plain of clouds resembling a frozen arctic sea, so that the successful climbers looked out upon a weird and lifeless world. The mood did not tempt them to remain long on the summit; a brief hour was sufficient, then they began the descent. Darkness caught them below the Cowlitz Cleaver and it was 10:30 that evening when they trudged into Camp of the Clouds to receive "three cheers and a tiger" from less adventurous companions.

The Northern Pacific ascent of Mount Rainier originated in obviously commercial motives, and yet, Olin D. Wheeler's discerning narrative was better than mere publicity; it was powerful support for the developing fight to establish a national park at the mountain. The accompanying three-color map, prepared by Henry M. Sarvent and G. F. Evans, was also the best available map of Mount Rainier prior to publication of the Geological Survey's topographic map in 1915 (and also first to show Fay Peak, honoring Washington's pioneering woman mountaineer) .[88] This ascent may also have encouraged the development of tourist facilities at Paradise Park, for Ross Comstock's participation in it led directly to the opening of "Paradise Hotel"— a tent camp on Theosophy Ridge — the following summer.

181

Turning away from the mountain proper, that summer saw another attempt to push through the legislation seeking establishment of a national park at Mount Rainier.

A third bill proposing a Washington National Park was introduced in the Senate on July 10, 1894.[89] Through it, Senator Squire attempted to appease the opposition by drastically reducing the area proposed for inclusion in the park (the distance between the northern and southern boundaries was reduced and the eastern boundary was made to coincide with the summit of the Cascades). The bill also moved the western boundary away from the mountain one township in an effort to correct the worst defect of the Pacific Forest Reserve proclamation. An area nearly identical to the present park was thus proposed.

Within a few days S. 2204 received powerful support, for Senator Squire was able to present the memorial to Congress from the joint committee of scientific societies on the 26th. Addressed *"To the Senate and House of Representatives of the United States of America in Congress assembled,"* the memorial first detailed the committees appointed for its preparation (a veritable honor roll of conservation), then proceeded to submit a statement of "the substantial reasons" for establishment of a national park to include Mount Rainier.

The memorial is well written, showing the detailed, accurate knowledge of Mount Rainier obtained by Bailey Willis in the course of his labors along the west slope of the mountain. We need not examine the text in detail here, for it is readily available,[90] but it is well to note the wisdom of those memorialists. They were aware that the western boundary of the Pacific Forest Reserve had been established too far to the east and failed to include the lower reaches of the western glaciers, and that Mount Rainier was situated well to the west of the summit of the Cascades so that the park need not extend beyond the crest of the range. Here is the outline of the park as they saw it:

> Your memorialists therefore pray that the park be defined by the following boundaries: Beginning at the northwest corner of section 19, T.18N., R.7E., of the Willamette meridian; thence south 24 miles more or less to the southwest corner of section

18, T.14N., R.7E.; thence east 27 miles more or less to the summit of the Cascade Range; thence in a northerly direction to a point east of the place of beginning.

Senator George F. Edmunds of Vermont, remembering his trip to the northwest slope of Mount Rainier with Bailey Willis in 1883, took the floor to add his testimony. He said: "I would be willing to go 500 miles again to see that scene. This continent is yet in ignorance of the existence of what will be one of the grandest show places, as well as a sanitarium."[91]

How much better it would have been had Congress seen fit to establish the park initially in accordance with the generous boundaries outlined by the memorialists. There would have been more "elbow room" north, west and south, and no need for a Mather memorial addition at a later date.

Some additional momentum was added on August 9, when Representative Doolittle presented the House with a petition from the Geological Society of America, asking the establishment of a national park to include Mount Rainier, and with a memorial from the Seattle Chamber of Commerce, requesting a survey of the public lands in the State of Washington.[92] After that, nothing more was done until the latter part of December, when Senator Squire presented a petition on which forty-nine members of the faculty of the University of Michigan signed their names in support of S. 2204.[93]

The year 1894 closed with an amusing episode which sent Major E. S. Ingraham off on an attempted winter ascent of Mount Rainier. It was a sufficiently audacious undertaking to be interesting despite its inevitable failure.

It appears that Henry Surry, the day jailer at the Seattle city jail, took a casual glance at Mount Rainier while riding to work on the Yesler Avenue car about 6:20 on the morning of November 21, 1894. He was astounded to see what he considered puffs of smoke coming from the top of the mountain. On his arrival at the jail, he mentioned that observation to patrol driver Frank Johnston. Johnston ran downstairs and told Charles Frasch, a grocer on Third Avenue, who went home for his opera glasses.

With the aid of the lenses, all agreed a catastrophy had

occurred on the summit of Mount Rainier. A volcanic eruption had most certainly taken place, for, as evidence, there was the smoke, and the discolored snow near the summit; yes, Columbia Crest was certainly gone, leaving the old mountain's top riven by a mighty fissure. After cutting in two more witnesses, the excited observers called a newspaper.

Worsening of the visibility about eight o'clock prevented a check by the staff, so members of Major Ingraham's party of the preceding summer were asked for their views on the matter. Dr. L. M. Lessey would not hazard an opinion; H. E. Holmes thought a tremendous avalanche had occurred; and Major Ingraham prefaced a wordy discourse by saying, "I hardly believe that there has been any great amount of change." That, and the immediate ridicule resulting from a dispatch to Tacoma, should have killed the story, but not so.

The perverse fate which orders geophysical events took a hand. At 6:30 that evening Tacoma was jarred by "several slight shocks of earthquake"; thereafter, it was easy to find people who could vouch for the fact that Mount Rainier had blown its top. Consequently, the story was used, and for all it was worth![94] The public reaction was quite similar to more recent hysterias concerning "flying saucers" and windshield pitting from "atomic fallout."

Such a furor was created that the *Post Intelligencer* decided to sponsor a winter ascent to determine what had happened. Of course, Major Ingraham was induced to lead the party of five, which included Dr. L. M. Lessey; George Russel, former Seattle postmaster; W. N. Sheffield, staff photographer; and two guides, E. Coke Hill and R. H. Boyd.

Leaving Seattle on December 17th, the party picked up some homing pigeons at Puyallup, then went to Wilkeson where they began their trek toward the mountain the following day. The attempt was made on snowshoes by way of the Bailey Willis trail to Spray Park, then across Carbon Glacier, and through Moraine Park to Winthrop Glacier, with the supplies and camp gear hauled on sledges. As they approached Winthrop Glacier on the day before Christmas, what appeared to be smoke and steam was seen coming from the summit of the mountain. The emissions were convincing enough to satisfy them an eruption *had* taken place; the purpose of the expedition was fulfilled, so they turned back the following day at a point about 9,000 feet in elevation, just below Steamboat Point.[95] Some good newspaper copy and about 100 photographs were brought back as a result of their adventurous winter mountaineering, but mountain climbers of the following season were unable to substantiate the conclusion regarding an eruption.

CHAPTER VIII

A NATIONAL PARK FOREVER

1895-1899

Though the popular movement to establish a national park at Mount Rainier quickly subsided, the idea received powerful and effective support. The work remaining to be done lay principally in the halls of Congress — in the preparation, initiation, and guidance of legislation intended to accomplish the grand objective for which the scientific societies had petitioned Congress. Even so, it required some time and some doing to bring about a successful conclusion.

PRESSURE TO OBTAIN FAVORABLE ACTION on Senator Squire's bill (S. 2204), was continued into the new year. On January 9, 1895, Senator McMillan, of Michigan, presented a petition signed by Professor W. J. Beal and other members of the Michigan Academy of Sciences, and Senator Squire was back ten days later with a petition from Wesleyan University — signed by the president, several members of the board of trustees, and twenty-five of the faculty. Again, on the 24th, Squire presented a petition from the faculty of the University of Wisconsin, at Madison, and the last petition, presented February 8, was from the president and nine professors of the University of Washington. All those documents were referred to the Committee on Public Lands, but to no avail.[1]

The summer of 1895 saw two tent camps established at Paradise Park. As soon as the snow permitted, Harry C. Comstock set up his "Paradise Hotel" on Theosophy Ridge, and a competitor by the name of Captain Skinner went

186

into business near him later. The first season was a break-even operation for Comstock, so he lost interest, selling out to John Reese the following spring.[2] Skinner was more persistent, operating his tent camp for several seasons before joining the gold rush to Alaska.

It was also during 1895 that the second Mount Rainier guidebook appeared. At the prompting of his many climbing companions, Major Ingraham published a forty-eight page booklet which was not only larger than Plummer's but more complete.[3] It contained the essential information on routes and distances, available accommodations and points of interest, several pages on the mountain's early history, a section on the flora (prepared by Professor Charles V. Piper), and some pointers for mountaineers — all of it well-illustrated with photographs and line cuts. Actually, it was but a short step from *The Pacific Forest Reserve and Mt. Rainier* to the popular guidebook sold by Barnes for many years.

Toward the end of 1895, Senator Squire and Representative Doolittle, of Washington, again introduced bills into Congress (S. 164 and H. R. 327) with the same purpose in view as before: establishment of a Washington National Park which would include Mount Rainier.[4] As before, it was the senator who led off, but this time he was not supported, for Representative Doolittle reverted to the text of his original bill, H. R. 4989, thus ignoring both the improved legislation proposed by Squire and the recommendations of the memorialists. Regardless of his reason— whether careless oversight or deliberate non-cooperation— Doolittle's action again drew an objection from the Commissioner of the General Land Office, and neither bill was reported back from committee.

When Representative Doolittle realized his bill would not be approved, he prepared another, H. R. 4058, which was presented on January 15, 1896. It was in essential agreement with Senator Squire's bill, and also offered a solution to the problem of the Northern Pacific Railroad grant lands lying within the proposed national park. The means hit upon to quiet the formidable opposition of the railroad was to offer it a patent on an equal amount of

187

non-mineral public land lying in any of the states through which its lines passed.[5]

On May 11th, H. R. 4058 was reported back from the Committee on Public Lands, with a recommendation that it be allowed to pass with amendments which would allow railroad and tramway rights-of-way within the proposed park, and make the provisions of the mineral land laws applicable within it. At last there was some hope for supporters of the measure and they must have watched with growing suspense as it progressed through the legislative toils.

Nearly a month later, on June 10th, Doolittle took the floor of the House to move that the rules be suspended to allow consideration of his bill. That was done, and Representative Bailey, of Texas, immediately voiced his objection to the measure. His views were dictated by a desire for federal economy, which he expressed as follows:

> I thoroughly sympathize in the desire of the gentleman [Representative Doolittle] to obtain this land for his State, and I am more than willing that he should have double the amount, but . . . I object to the Secretary of the Interior becoming the keeper of a park; and I object to taxing all of the people of the United States to maintain a park in the State of Washington. If the gentleman will prepare an amendment that gives this land absolutely to the State of Washington, I will support it.[6]

Doolittle was so eager to obtain favorable consideration of his bill that he answered Bailey in a manner that would embarrass supporters of later legislative efforts, saying:

> You understand that we are not asking for an appropriation, and do not intend to ask for a nickel from this government. We propose to make the necessary improvements, for the benefit of all the people of this country, out of our own pockets, every dollar of it, and we are asking nothing in the way of appropriation. It seems to me that no fairer proposition has ever been made to the Congress than this.

With the wry comment that his difficulty was that he had "not learned how it is possible to maintain a park by any government without expense," the Texan dropped the argument.[7]

Amended in title, the bill passed the House then went

188

to the Senate and the Committee on Forest Reservations and Protection of Game the following day. There it remained until February 17, 1897, emerging without further amendments, to obtain a late place on the Senate calendar.

When the bill did come up for consideration in the Senate, on March 3rd, its passage was delayed by the argument of Senator Vest (the self-appointed guardian of Yellowstone National Park), who felt there was no place for railroads or tramways in any national park. The bill to create a Washington National Park did pass, but not in time to receive the President's signature before he vacated his office on the following day.[8] It was a great disappointment.

Though President Grover Cleveland failed to sign the bill which would have created a national park at Mount Rainier, he did change the name of the area from the Pacific Forest Reserve to the Mount Rainier Forest Reserve by a proclamation dated February 22, 1897.[9]

The year 1896 cannot be passed without taking note of two interesting ascents of Mount Rainier. The Tacoma violinist, Olof Bull, finally reached the top of the mountain with his violin on July 28th. There he indulged his whim to give a concert on Columbia Crest, ending it, quite appropriately, with "Nearer, My God, to Thee." But his diary is of greater interest than his musical stunting, for he chanced to record a remarkable solo climb which may have been the first such ascent.[10] He noted it thus:

Monday, July 27. Arose early, and with A. H. Waite, H. B. Dewey and Robert Newell, started for the top of the mountain at 5 o'clock. On the way were passed by two gentlemen, Peterson and Bergh, and then A. B. Wood came along, having left our camp at 6:15, and marched defiantly by everybody, scaling the loftiest peak of the mountain at 12:50, making the trip without a single stop. He remained at the summit until 2:20, making a thorough investigation of the crater, then retraced his steps to the valley, reaching camp at 6:15 p.m., having accomplished the wonderful feat of reaching Columbia Crest from Theosophy Ridge and return in 12 hours' time.

The other ascent was less spectacular though far more important. Bailey Willis entered the United States Geo-

189

logical Survey after his employment with the Northern Pacific Railroad was so abruptly terminated, and he rose rapidly in that service. However, his success was gained at the expense of his health, which became precarious — consumption, according to his doctor.

That led Bailey Willis to ask for a field assignment in the hope that it would improve his health, and the summer of 1896 found him again in western Washington assisting with the survey of the Tacoma quadrangle. But he proved an odd sort of invalid, for he soon joined a second cousin, Israel C. Russell, who was a professor of geology at the University of Michigan, in organizing a party for geological exploration at Mount Rainier. The group which left for the mountain included little Hope, the ten-year-old daughter of Bailey Willis, Professor Henry Landes, George Otis Smith, William B. Williams, a Swede by the name of F. H. Ainsworth, and a lame French cook, Michael Autier.[11]

They traveled over Bailey Willis' old trail as far as the Carbon River crossing, then followed a newly-opened path up the north bank of the river to Chenuis Falls. From there they cut a way to the foot of Carbon Glacier, established a supply depot and pushed on up the lateral moraine to a beautiful camp in the alpine meadows between Carbon and Winthrop glaciers.[12]

That camp was the base for an ascent of the mountain, and there little Hope was left with the lame cook and Professor Landes, who was unwell. Though Russell "repeatedly said she could go to the top of the mountain as well as any," Bailey Willis feared that the physical strain and the notoriety of a successful ascent would both be too much for his daughter; so he kissed her and followed the other climbers upward, leaving a brave little girl to a lonely vigil.

After passing over the ridge between the base camp and Winthrop Glacier, the climbers took to the ice and roped up, for Russell, the conqueror of Mount St. Elias, was no amateur. The order of march up the glacier was Russell, Smith, Willis, Ainsworth, and Williams, with "experience being at one end and strength at the other." The smooth, white center of the ice flow appeared to offer an easy route

190

toward the rocky point of The Wedge, where they intended to make a high camp, but they were deceived, for there were many hidden crevasses requiring frequent detours. Actually, seven hours of hard work passed before they could throw their blankets down on the rocks of Ingraham's "Battleship Prow."

Tea was made on an alcohol stove, a supper of hard tack, ham, and chocolate was quickly eaten, then each man built himself a little sleeping terrace among the loose rocks below the cliff. In a letter to his mother, Bailey Willis said:

> The night was cloudless and moonlit and I lay awake a long time watching the scene, but the sharp frost drove me beneath my blanket. About half past four I turned out to light the "Baby" as Russell had dubbed the alcohol lamp. The sun was still behind the smoke bank which enveloped the entire landscape, and formed a horizon of deep copper color; but behind me the great peak was brilliant with golden light.

Leaving their blankets at that high camp, the climb was resumed up the "wildly crevassed ice cascades" toward the summit, which they hoped to reach in time to descend that day. Russell picked his way confidently among the seracs and crevasses with the caution of a competent mountaineer; always "quiet and self contained, his manner never betrayed any excitement." When a snowbridge had to be crossed, he would tell Smith to plant his alpenstock and take a turn of the rope about it, then he would cross over and give Smith the same security.

Bailey Willis admits that "in reading of accidents to parties roped together I had thought I would like to have such an experience if it might end happily — and I have had my wish."

> We were on a snow slope, steep as a gothic cathedral roof. Below was a wide and deep crevasse. Russell and Smith had gone to the length of the rope along the slope, I had followed and we three had anchored ourselves by driving our alpenstocks deep in the snow—the two men behind were told to follow. Ainsworth took a step and slipped. He dragged Williams down and with a little cry of "Oh" Williams disappeared in the crevasse, while Ainsworth clutched at its edge. The jerk came on me and nearly cut my waist in two, but I hung on to my alpenstock and it held, supported partly by Smith's hold. In a

191

moment Williams appeared climbing up the rope behind Ainsworth who drew himself out on the snow. Then came two or three quiet orders from Russell, and the episode was over. Williams, however, had dropped his alpenstock and as he, the end man, must have one, I, the middle man made the rest of the trip without one.[13]

Russell is said to have been unwilling to continue the climb with men who could not stay on their feet, but the others overruled him and the ascent was continued.[14]

For five hours Russell led us upward and diagonally across the glacier, repeatedly approaching and retreating from a zone of huge crevasses which he wished to cross to reach a smooth slope that led temptingly to the summit. At last he took a long survey from a commanding ice pinnacle and turning back said quietly but finally: "It can't be done." There was, however, another possible way, more to the north and we decided to try it. Three hours later Russell cast off the rope at the base of the cone of loose rocks which forms the craters outer rim and climbed on ahead. Not one of us could follow him closely—The wind blew a gale. Exhausted for breath we would take a few steps and drop on the rocks. At last I looked over the crest into the bowl shaped crater half a mile in diameter—it was filled almost level with snow and the wind howled across it. But Russell called to me from a crevice between the snow and the rocks and descending to him I found a cavernous gallery whose roof was ice and whose walls were hot rocks emitting clouds of steam.

It was four o'clock on the afternoon of July 24, 1896, and obviously too late to descend by the perilous route they had come up, particularly in view of the physical condition of the climbers. Only Russell was capable of sustained exertion, while Williams and Ainsworth were overcome by headache and nausea; an overnight stay on the summit was clearly a necessity.

Supper consisted of tea, with a hardtack and a bite of ham apiece, for that was all they had. A rope was then stretched along the steaming wall of the crater to serve as a hand line, rubber raincoats were donned and they were ready for the night. Several of the climbers moved about restlessly during the dark hours, but Bailey Willis "lay in one place, taking now the breath of hot steam, now the blast of freezing wind, and occasionally shifting the pressure of the sharp rocks." For him the night wasn't particularly long.

At daylight some icicles were melted to make tea, then they mustered courage to face the wind and climb to the top of Columbia Crest where Bailey Willis waved Hope's small silk flag and Ainsworth planted his McKinley button "on the highest point in the U. S. south of Alaska."

After a brief look over the sea of smoke, from which a few peaks projected, the party began the descent by the Gibraltar route to Paradise Park, which they reached at two o'clock that afternoon. The weary, hungry climbers staggered into the camp of that old veteran, Major Ingraham, who was vacationing there, and he did all in his power to assist them. Out of gratitude, the geologists later named a branch of the Cowlitz Glacier in honor of Edward Sturgis Ingraham.[15]

With provisions for one day, which they deemed sufficient for the return to their base camp, the climbers set off across the glaciers lying on the mountain's east flank. But they miscalculated, for night found them on the north side of the Cowlitz Glacier, bivouacking in a clump of stunted pines "near a charming cascade."

Bailey Willis says they were more fortunate on the fifth day, so that the "ways seemed to open before Russell and we attained one objective point after another until towards sunset I looked down upon the flag of our camp and could see Hope sitting occupied by the fire. She was sewing."

The little girl's vigil at "Star-lily Camp" had not been an easy one. She had kept busy gathering firewood, collecting flowers, mending, and floating little boats carved for her by the kindly Autier. "It was most lonely at night," she told her father afterward, "with the others asleep out there. I don't know what I would have done without Jacky, [a rubber dog] I would squeeze him and he would squeak and it was some company."[16]

Looking back over the intervening fifty-eight years, Mrs. Rathbun (who was once Hope Willis) says:

> I think those last two days, when we thought they were lost, must have been rather terrible to me but I only remember that final Friday evening. The mountain stood very clear in the

sunset and after a frugal supper (I am sure I was given the best of it) Professor Landes sat whittling by the fire. He'd been nearly silent all day, and I could bear it no longer and asked "Do you think they'll come tonight?" He answered with irritation, "If they don't, it will mean a search for them tomorrow." I can still remember how my heart went down and the vague wonder at how the sick man and the lame M. Autier could ever find them in the vastness above us. But then came Father's "oopee," loud and cheery, from the slopes above us and in a few minutes Father came swinging into camp ahead of the others.

The climbers were scarecrows, "with hardly a whole garment among them . . . their lips black and their hands and arms swollen." Though they were safely back, ending the loneliness of a little girl and banishing the terrible fears of a sick geologist and a lame cook, their trials were not ended. The party's supplies were nearly exhausted and they had to leave the mountain in haste. The camp of the packer, Fred Koch, who waited for them at the end of the trail, was no better off, but from there it was only a day's ride over a trail threatened by a forest fire to Carbonado and the end of an adventure.

Rugged as the experience was, the consumptive who broke his cod liver oil bottle on the trail to the mountain, before making an arduous ascent in summer clothes, found it good medicine. He was not troubled further, and lived a long and useful life to the age of ninety-two.[17] Bailey Willis returned to Washington, D. C., with notes and photographs of inestimable value to the continuing effort to establish a national park at Mount Rainier; while Israel C. Russell and George O. Smith both wrote scholarly reports — one on the mountain's glaciers and the other on its rocks — thereby laying the foundation of our knowledge of Mount Rainier's geology.

A club-sponsored outing, which featured an ascent of Mount Rainier, was the novelty of the summer of 1897. It began on the morning of July 19th, when a large group of Mazamas left Portland by train for Tacoma where they were honored at a city hall reception. The following morning, the dusty trip to the mountain was begun in horse-drawn stages which took them to Eatonville the first night, Kernahan's ranch the second, and to the end of the road by noon of the third day. Most stayed over at Longmire's Springs until the next day, but a few pushed on up the trail to the alpine meadows where they were hospitably received at a small camp already established on Magnetic Ridge by Seattle and Walla Walla members of the club.

Clouds closed in on the 23rd, bringing rain to dampen the enthusiasm of those who straggled up from the springs that day. Nor was it cheering to find that one of the earlier arrivals, Professor J. E. Brown of Stanford University, had foolishly started alone for the summit of the mountain at 2 a.m.—just ahead of the developing storm. At ten o'clock that night a six-man party was sent out to search for the imprudent climber, who was found crouching in the lee of a boulder less than two miles from camp.

The following stormy day was spent putting "Camp Mazama" in order. Forty-five sleeping tents had to be put up to shelter more than 200 club members and their friends, four tons of supplies had to be brought in to a safe storage, and pasturage had to be arranged for two beef steers, seven milk cows and many horses.

Saturday, the 24th, continued cloudy but was a good enough day for a party of twenty to make a conditioning climb on Pinnacle Peak. However, it was Monday before serious mountaineering began. Early that morning four climbers started for the summit with the intention of staying overnight in the crater, and they were soon followed by eight more who proposed to make the top and return that day (which five succeeded in doing). At 9 a.m. the main party left the base camp, climbing leisurely to Camp Muir, which a pack train had already stocked with food and bedding.

After the usual sleepless night, the hopeful climbers were marshalled in the early dawn to begin the ascent in accordance with a pre-arranged plan by which they were divided into four groups, each under a captain. The individual climbers were required to "keep in line, to obey orders implicitly and to use no stimulants while climbing," and it was further stipulated that ladies must wear bloomers.[18]

The ascent was a slow one, with some climbers dropping

195

out on the grind up the Cowlitz Cleaver and others at the fearsome Gibraltar ledge. Above it, steps had to be cut for 1,000 feet, so that it was 12:20 p.m. before Camp No Camp (now Camp Comfort) was reached. Nor did matters go much better on the summit dome, where increasing fatigue slowed the progress of the unwieldy serpentine of climbers. As a result, the first detachment did not reach the crater rim until 3:30 on the afternoon of July 27, 1897.

In all, fifty-eight members of the club were registered on the summit as having completed the climb, among them Edgar McClure, Professor of Chemistry at the University of Oregon. For him, mountain climbing had become but a means to an end — measurement of the height of mountain tops. He had already determined the height of Diamond Peak and the middle peak of the Three Sisters, in Oregon, and of Mount Adams in Washington. The Mazama outing provided him with a chance to do the same for Mount Rainier.

Thus, McClure carried his favorite instrument, a yard-long mercurial barometer, to the top of the mountain, obtaining with it, a summit observation of 17.708 inches of mercury at 4:30 p.m., Pacific standard time. The hour is important.

Leaving eight climbers to spend the night in the steam caves (among them Fay Fuller, to whom it was no novelty), the unwieldy main party began the descent at 4:40 p.m. At the Gibraltar ledge there was a near tragedy when J. F. Anderson lost his footing and dragged another climber off the narrow, sloping shelf. Fortunately, the "life-line" they were grasping held and they were able to scramble back to safety. Without further incident, the last of the weary, and in some cases, nearly exhausted climbers reached Camp Muir at 9:30 p.m., which was well after dark!

There a fatal mistake was made. Mr. E. S. Curtis, who had been appointed to marshal the ascent, relaxed the strict control which had been the chief safeguard for those inexperienced on Mount Rainier. About half the climbers decided to stay overnight at Camp Muir, while the others broke into several parties for the trek to Camp Mazama. The first group consisted of five climbers, with Professor

196

McClure in the lead — still carrying his barometer, but without a light.

Traveling in darkness, largely by the feel of the terrain, he inadvertently stepped over the edge of what is now known as McClure Rock, sliding to his death on a rocky outcrop below.[19] The pioneering early ascents — those dangerous scrambles on largely unknown heights — were well in the past that evening of July 27, 1897, when the mountain's first climbing tragedy occurred.

McClure's death was not the only accident to plague the Mazama outing. The party which remained overnight on the summit appears to have broken up at Camp Muir on the following day, with two Portland men, H. C. Ansley and Walter Rogers, descending from there on the evening of the 28th. About 200 yards east of the point where Mc-Clure was killed, they slid down a snow slope out of control to plummet into a crevasse. Mr. Ansley extricated himself and came into Camp Mazama "cut and bleeding and nearly crazed with fatigue and excitement," bringing word that Rogers was still in the crevasse. A relief party left for the scene of the accident at midnight and soon located the unfortunate climber, who was "still conscious, though nearly frozen." He begged for help, "as many had merely looked in on him and gone away without a word."[20] A rescue was soon effected by means of the ropes which the relief party brought up.

The victims of the second accident were luckier than McClure, for they recovered, but the cumulative effect of misadventure was to shorten the Mazama outing considerably. The club members were back in Tacoma at 3:30 p.m. on August 3rd, concluding an outing which might have lasted a fortnight longer under happier circumstances.

September saw the passing of another mountaineer. James Longmire, pioneer of 1853, veteran of the Indian war, explorer of the wilderness about Mount Rainier, and proprietor of Longmire's Springs, died during the month at the home of his son, Robert, at Tacoma.

During the thirteen years from the time he located at the springs until his death, the energetic old man had brought his vision of a tourist Mecca at Mount Rainier to a

197

homespun reality. His improvements included the erection of an "inn, barn, two bath houses, one storehouse, and two small shacks,"[21] construction of nine miles of wagon road over which stages operated during July, August and September of each year after 1896, and the building and maintenance of trails which made Paradise Park available to the public. It was a great work.

The effort to push legislation for the establishment of a national park at Mount Rainier was never allowed to lag. In the fifty-fifth Congress new men took over the work which had been pushed so persistently by Senator Squire and Representative Doolittle. President McKinley was hardly settled in office when Senator John L. Wilson (Washington) introduced S. 349, a duplicate of the almost successful bill of the previous session, on March 16, 1897. His colleague, Representative Jones, introduced a similar bill, H. R. 5024, on December 15. Neither bill was reported back from committee, but an identical one, S. 2552, introduced by Wilson on December 7 would finally win through.[22]

With the fate of Wilson's bill yet uncertain, Representative James Hamilton Lewis introduced another measure proposing establishment of a Washington National Park at Mount Rainier,[23] but it, too, joined the long roll of such bills to die in committee.

While Wilson's bill, S. 2552, was in the hands of the Committee on Public Lands with its fate undecided, supporters of the national park movement brought to Washington a man whose influence assured its passage in the Senate. After his return to the East, the Reverend E. C. Smith gave a lecture on Mount Rainier at a meeting of the Appalachian Club in Boston. It was illustrated with lantern slides made from some of W. O. Amsden's photographs taken at the mountain in 1890, and both the lecture and the well-written article subsequently published in *Appalachia* gave Smith standing as an authority on Mount Rainier. Thus, he was one of the men chosen to represent the Appalachian Club in the preparation of the great memorial presented to Congress in 1894. He says:

It was on the strength of this article in *Appalachia* and particularly, I judge, on account of these pictures [Amsden's] that I was invited to lecture before the National Geographic Society in Washington when the bill to create the national park was being discussed in Congress. I was in Washington four days in which I had several meetings with Bailey Willis of the Geological Survey, who had recently done some work in the Puyallup and Mowich glacier areas. He arranged for additional slides from the Amsden photographs and for a map of the glaciers taken from a German atlas. It was quite incorrect, but impressive and much to our purpose.

At a dinner at Gardner Hubbard's home [president of the National Geographic Society] I met the Pacific Coast senators and officers of the Society and answered many questions about Mount Rainier. The Society put its influence strongly in support of the bill . . . Not long afterwards a vote was taken and Mt. Rainier became the name and the commanding feature of a new national park.[24]

That meeting in the home of Gardner Hubbard created a favorable climate for the passage of the Mount Rainier legislation in the Senate, and the next day, on June 9, 1898, it was sent to the House and referred to the Committee on Public Lands. But there, powerful opposition developed.

Representative Lewis, a Democrat, was piloting the bill through the House when it developed that the conservative influence of "Uncle Joe" Cannon, a Republican from Illinois and chairman of the House Ways and Means Committee, would probably prevent its passage. Colonel Lewis hastily called John P. Hartman, who had been largely responsible for framing the bill, to the capitol early in February, 1899, and together they called upon Chairman Cannon. Hartman has described the meeting thus:

As usual, Mr. Cannon was smoking his big, black cigar, ensconced in a swivel chair, with his feet on the jamb above the little fireplace where coal was burning cheerily in the grate.

After preliminaries Mr. Cannon said, addressing me, "I have a notion to kill your Bill, and I have the power to do it." Of course I wanted to know the reasons and he said: "It is all right to set these places aside but for the fact that in a year or so you will be coming back here seeking money from the Treasury to improve the place, and make it possible for visitors to go there, which things we do not need, and we haven't the money therefor, and I think I will kill it." I said to Mr. Cannon, "I promise you, Sir, that if the Bill is passed I will not be here

199

asking for money from the Federal Treasury to operate the place so long as you shall remain in Congress." With that statement he said, "I will take you at your word and let the measure go through, if otherwise it can travel the thorny road."

I kept my word sacredly with Uncle Joe and never did I by voice or otherwise ask one cent of appropriation for that Park during the time Mr. Cannon remained in Congress, but others did and some millions were appropriated and all, I think without his protest, for he saw the light later on, and realized that his view and conservatism was not the right principle for so great a country.[25]

Representative Lewis was successful in getting the bill out of committee on February 24 (with an amendment which gave settlers on lands within the proposed park the right to seek other public lands in lieu of the holdings they would lose); then, on March 1st, Representative Lacey, of Iowa, called up the bill for consideration of the House. After the bill was read, Mr. Lacey asked that the name of the area be changed by amendment from "Washington" to "Mount Rainier"— the amendment was written in and the measure passed. It was immediately sent back to the Senate for the concurrence of that body. The Commissioner of the General Land Office made a favorable report on S. 2552 on March 2, 1899, and it was signed by President McKinley the same day to become the organic act of Mount Rainier National Park.[26]

And so, through the efforts of men who had the mountain fever, a new jewel was added to the nation's heritage. It had a new park — its fourth — and the first founded on the theme of mountain grandeur, in a system which is distinctively American. May the mountain remain "a national park forever."

EPILOGUE

The early history of Mount Rainier logically ends in the year 1899, with enactment of legislation conferring a national park status; and yet, that would leave some of the story untold.

For a time, the area was no better off as a national park than it had been when a part of the reserved lands. Its affairs were administered by Forest Supervisor Grenville F. Allen (son of the retired professor who settled in Succotash Valley in 1889) from his office at Orting, Washington, and, as the acting superintendent, he probably did little more before 1903 than make an occasional trip to the park—which is understandable, since he had no appropriation for its care.

Actual development did not begin until the park was four years old. Representative Francis W. Cushman was able to get an appropriation for $10,000 for the fiscal year 1903, for construction of a government road from the end of the Pierce County road, near Ashford, to Paradise Park. The survey was begun July 11, 1903, by a crew working under the direction of Oscar A. Piper.[1] As originally laid out, the center line of the road passed directly over the Hudson grave, but the stakes were moved enough to clear it when the Longmires brought the matter to the attention of Eugene Ricksecker, the civilian engineer in charge of the project. Once started, the work proceeded rapidly and the road was roughed out as far as the crossing of Nisqually River by 1905.

Among those who rode the cars of the Tacoma and Eastern Railroad to its new terminus at Ashford, and followed the new road to its raw ending below the glacier, was General Hazard Stevens. He was returning, after an

201

absence of thirty-five years, to make a second ascent of the mountain.

The old mountaineer noted many changes on the upper slopes, making particular mention of the following:

> The ledge or path along the side of Gibraltar has changed considerably. It is not so narrow and well defined, the talus is wider and larger, the upper strata having crumbled and fallen down more than the lower ones . . . On the whole this part of the ascent did not seem so difficult or dangerous as it was in 1870.
>
> I am satisfied that there was much more snow on the summit in August, 1870, than last July. Then the snow completely filled the western, or smaller crater, leaving only a narrow rim of rocks showing above it. Now the snow was 40 feet below the rim. The ice cave in which we took refuge had disappeared and only a feeble emission of steam and heat remained . . . But most surprising change of all, the wind-swept rocky ridge of 1870 separating the two craters, where I deposited the brass plate inscribed with our names and date, August 17, 1870, is now covered by a large, regular mound of snow 30 feet high.[2]

In regard to the appearance of the summit, it is obvious that Stevens' memory tricked him, for the account he published in *Atlantic Monthly* in 1876 indicates the ice hummock we now call Columbia Crest was there in 1870.[3] However, it may have been smaller, afterwards building over the rocks where Stevens deposited the tokens of conquest. Some day, when the conditions of 1870 are reproduced on the summit of Mount Rainier, the brass plate and the old army canteen left in the cleft of a boulder on what Stevens thought was the highest point may again be seen.

The following year, 1906, the first park ranger went on duty at Mount Rainier. (Two forest rangers, William McCullough and Alfred Conrad, had been assigned to cover the Nisqually and Carbon River portions of the park after 1903.) He is familiar for he was the same Oscar Brown who raised the flagpole on the summit in 1891. His first concern was to build a ranger station, which he did with the aid of a horse to skid the logs and raise them. The little building remains to this day, standing just inside the Nisqually gate on the right hand side of the road, inbound, having served as ranger station, superintendent's headquarters, and residence in the half-century since it became the first government building in the park. It is easily identified for the fan-like ornamental logwork in the gable is unique.

Four years later the government road was completed to Paradise Park at a total cost of $240,000 for the initial construction[4]—a considerably higher figure than the $15,000 estimated as sufficient by Plummer in 1892. Yet the road was too poor for automobile and motorcycle traffic above the Nisqually River bridge, and it was not until August 8, 1911, that the first car, with President Taft as a passenger, reached Paradise Park — dragged through the mud by a team of mules from Narada Falls on up! Actually general automobile traffic was not permitted above the bridge until June 20, 1915, and then only under a strict traffic control system.[5]

Another dream of yesteryear became a reality in 1915 with the completion of a trail encircling Mount Rainier.[6] Now we know it as the Wonderland Trail.

While passing rapidly over these later years, that valiant old mountaineer, P. B. Van Trump has been overlooked. Van Trump was a cultured but impractical man who never made a really satisfactory compromise with the drudgery of earning a living. He had an adventurous nature and a scholarly mind, and he was a consummate cook, but there his talents ended. It was his wife, Cynthia, who ran the general store at Yelm, and she did all the farming done on their place. An old neighbor has characterized him with the statement, "Mr. Van Trump was a fine old gentleman, but he never learned how to work." As a result, he did not prosper with the years.

After his wife became sickly, Van Trump moved to Seattle where he and his daughter, Christine, ran a boarding house.[7] He lost both his wife and daughter in 1907, leaving him homeless and alone. An old friend, Mrs. Susan Hall, opened the Wigwam camp at Indian Henry's in 1908, and she asked him to come and help her. It was there the kindly old man endeared himself to a generation of tourists by his solicitude for their welfare, and it was there he finally became a seasonal ranger. Probably it was one of the glories of his life that he was able to wear the badge of a federal officer in the park he had done so much to bring into being.

Thus were the last years of his life passed — summering at the mountain and wintering with the Longmires at Yelm — until the day in 1915 when a reporter wrote a column head, "No Mountain Trip This Year for Van Trump."[8] It could not have been an easy decision for a man who loved Mount Rainier as he did, but his seventy-six years lay heavily upon him and he felt his vigor ebbing.

Van Trump returned to New York State to spend his remaining days with his people; and it was there, on December 27, 1916, that he "just fell asleep like a child tired out from play."[9] General Hazard Stevens, his old comrade and friend of many years, made the near perfect comment: "In his passing the world lost a splendid fellow."[10]

The files of the Yakima Indian Agency provided a letter which was the clue to an amusing and deeply symbolic story. It reads as follows:

September 13, 1915

United States Indian Agent
Fort Simcoe, Washington.
My Dear Sir:

The papers are telling of a Chief Sluiskin and 30 Yakima braves insisting on hunting privileges in the Mount Rainier National Park. To me, as a historian, the interesting item is the statement that Sluiskin who guided Stevens and Van Trump on their climb of Mount Rainier in 1870 is still alive. Will you please let me know if this is true, and if he is still alive, please let me know where his home is.

Yours faithfully,
Edmond S. Meany.

The story Professor Meany referred to appeared in a number of papers on September 10, relating an incident which probably occurred in the vicinity of Yakima Park toward the end of August.[11] Park Rangers Leonard Rosso and Arthur White came upon a band of thirty Indians encamped in two tepees, "eking out an existence in the manner of their kind before the coming of the white man"— that is by hunting.

The rangers found the leader of the band to be a wrinkled, old chief whose named sounded like "Sluiskin"

204

to them, though it probably was pronounced Saluskin. They explained, through a girl interpreter, that hunting was not allowed in the park, but to no avail; the old Indian merely went into a tepee and returned with three papers which he handed them, then stood by smiling and commenting in the Yakima tongue while the soiled documents were read. One proved to be a copy of the treaty Governor Stevens made with the Yakima Indians in 1855, and the others were open letters — one written by a police judge and the other by a former Indian agent — stating that the old chief should not be molested.[12]

Now this was not the first such case for the small ranger force of two permanent and five seasonal men, and a lack of support at the Washington level had left the park officers in some doubt regarding their jurisdiction. And so, the rangers withdrew, to report to Supervisor D. L. Reaburn, which took some time since they had no telephone. The supervisor telegraphed Washington, then sent Chief Ranger Thomas O'Farrell from his headquarters at Fairfax to find the Indians. But O'Farrell had his trip for nothing; the hunting party was gone from the park when he reached their campsite.

Without doubt the incident would have attracted as little attention as the earlier trespasses had,[13] except that the newspapers identified the old chief, Saluskin, with Sluiskin-of-the broken-hand, and proceeded to manufacture a real old tear jerker story. Here was the noble redman, deprived of the rights guaranteed him by treaty and hounded from his ancestral hunting grounds — a poor recompense for loyally guiding the first white men to scale the mountain. It was "Lo, the poor Indian," played well and loudly!

Actually, the Yakima claim to hunting privileges on the eastern flank of Mount Rainier was not very good. The reservation established by the treaty of 1855 did not include territory west of the Cascade Mountains, despite the fact that the Yakimas had made some use of certain west side areas for berry picking and hunting; nor could the special privileges they secured by the treaty be construed to apply after the park was withdrawn from the "open and unclaimed lands" of the United States.[14]

But that aspect of the incident was ignored in favor of its emotional appeal. The newspaper story connecting the old Indian with the Stevens-Van Trump ascent of Mount Rainier in 1870 was questioned at once by Joseph Stampfler, who pointed out that Sluiskin "was drowned in an Eastern Washington river about 10 years ago,"[15] an opinion supported by David Longmire three days later.

Soon after Professor Meany wrote his letter of inquiry to the Yakima Indian Agency, the story was further confused by a newspaper report which "confirmed" that here, indeed, was the Sluiskin of 1870. The information was attributed to A. J. Splawn, a former mayor of North Yakima and local historian,[16] and it was given further support when a lawyer by the name of Lucullus McWhorter interviewed Chief Saluskin and also misconstrued his narrative of the trip to Mount Rainier in 1855 when he guided the two unknown white men. By that time the facts were so clouded that even the newspapers lost interest.

Both Splawn and McWhorter later put the story straight, but, by then, few cared. The old chief died late in 1917,[17] as obscurely as he had lived, his life having spanned almost exactly the period from the founding of Nisqually House and the first interest of white men in Mount Rainier to the establishment of National Park Service administration over the mountain and its environs.

The year 1918 saw the passing of yet another of the principal characters in this drama. From 1874 to 1916 General Hazard Stevens lived in Boston, where he practiced law and cared for his widowed mother. After her death he returned to Washington State to settle near Olympia, in sight of the great peak he had conquered in 1870, and that was home to him until his sudden death October 11, 1918, at a pioneer reunion.

Many others have followed him, but not all. As I write, Fay Fuller (Mrs. Fritz von Brieson), Reverend Ernest C. Smith, and Leonard, Susan, and Maude Longmire, and perhaps some I am unaware of, are yet with us — living links with the days of Mountain Fever.[18]

APPENDIX

A SUMMARY OF KNOWN ASCENTS AND NEAR ASCENTS
OF MOUNT RAINIER
1852-94

(1) August 21,* 1852. See pages 10-13.

Party: Sidney S. Ford, Jr., Robert S. Bailey, John Edgar, and probably Benjamin K. Shaw.

Route: The mountain was approached by way of the Nisqually Valley, and the south side was ascended, perhaps to an elevation of 14,000 feet, with at least one overnight camp on the upper slopes.

These appear to have been the first white men to go above timber line on Mount Rainier.

References: Hubert Howe Bancroft, *History of Washington, Idaho, and Montana, 1845-1889* (San Francisco, 1890), 386.

George H. Himes, "Very Early Ascents," in *Steel Points*, I (July, 1907), 199-200.

"Visit to Mt. Ranier," in Olympia *Columbian*, September 18, 1852.

Leslie M. Scott. "News and Comments," in *Oregon Historical Quarterly*, XIX (December, 1918), 337-38.

(2) June, 1854.** See pages 15-18.

Party: Two unknown white men, guided to the mountain by Saluskin, a young Yakima Indian.

Route: The mountain was approached from the east, and a one-day ascent, via the Winthrop and Emmons glaciers, was made from an encampment at Mystic Lake, where the guide remained.

These appear to have been the first white men to reach the summit craters.

References: L. V. McWhorter, "Chief Sluiskin's True Narrative," in *Washington Historical Quarterly*, VIII (January, 1917), 96-101.

A. J. Splawn, *Ka-mi-akin, The Last Hero of the Yakimas* (Portland, Oregon, c. 1917), 343-45.

*The day is uncertain.
**The year is uncertain.

(3) July 10, 1857. See pages 21-28.

Party: Lt. August V. Kautz, Dr. Robert Orr Craig, and privates Nicholas Dogue and William Carroll, guided by Wa-pow-e-ty, an old Nisqually Indian.

Route: The mountain was approached by way of the Nisqually Valley to a high camp above the moraine of Nisqually Glacier, on the west side. From there, an ascent was attempted via the Kautz Glacier, with the lieutenant reaching an elevation of about 14,000 feet.

The Indian guide climbed without apparent qualms.

207

References: George Gibbs, "Physical Geography of the Northwestern Boundary of the United States," in *Bulletin of the American Geographical Society of New York—1872*, 356-57.

John S. Hittell, "Some Notes on the Resources, Population and Industry of Washington Territory," in *The Hesperian*, III (November, 1859), 398-99.

August V. Kautz, Ms. journal of Lt. August V. Kautz, 1857-61; Microfilm A-383, 8-12, the National Archives.

_____, "Expedition to the Summit of Mt. Ranier," in Steilacoom *Washington Republican*, July 24, 1857.

_____, "Ascent of Mount Rainier," in *Overland Monthly*, XIV (May, 1876), 393-403.

"He Reached the Top," in Tacoma *Ledger*, January 31, 1893.

(4) August 17-18, 1870. See pages 30-51.

Party: Hazard Stevens and Philemon Beecher Van Trump, guided to Bear Prairie by James Longmire, and from there to the mountain by Sluiskin, a Yakima Indian.

Route: The mountain was approached by way of the Nisqually Valley to Bear Prairie, then across the Tatoosh Range to a base camp near Sluiskin Falls. From there, the two white men made an ascent via the Gibraltar route to Success Peak and the principal summit, staying overnight in the small crater.

Edmund T. Coleman accompanied the party to Bear Prairie but did not climb the mountain. This is the first party to stay overnight in the summit steam caves, and the first to carry flags to the top. Van Trump was injured.

References: "Ascent of Mount Rainier," in Olympia *Tribune*, August 29, 1870 (reprinted in Olympia *Transcript*, September 3, 1870; Olympia *Weekly Echo*, September 3, 1870; Tacoma *Every Sunday*, October 15, 22 and 29, 1892, and Seattle *Post Intelligencer*, January 6, 1895).

Edmund T. Coleman, "The Ascent of Mount Rainier," in H. W. Bates' *Illustrated Travels*, V (London, n. d.), 161-67.

Hazard Stevens, "First Successful Ascent of Mount Rainier," in *Atlantic Monthly*, XXXVIII (November, 1876), 511-30.

P. B. Van Trump, "Mount Rainier," in *Mazama*, II (October, 1900), 1-18.

Laura Virginia Wagner, *Through Historic Years with Eliza Ferry Leary* (Seattle, 1934), 14.

(5) October 17, 1870. See pages 51-57.

Party: S. F. Emmons and A. D. Wilson, guided to the mountain by James Longmire.

Route: The mountain was approached by way of the Nisqually Valley, Skate Creek, and Backbone Ridge to a base camp near Little Tahoma. From there, the two surveyors crossed the Cowlitz Glacier to a high camp near the Cowlitz Rocks, and then made a one-day ascent to the summit via the Gibraltar route.

Wilson carried a theodolite to the summit but was unable to make use of it.

References: S. F. Emmons, "Glaciers of Mt. Rainier," in *American Journal of Sciences and Art*, CI (March, 1871), 157-67 (Vol. I, 3rd Ser.).

_____, "Ascent of Mt. Rainier, with Sketch of Summit," in *The Nation*, No. 595 (November 23, 1876), 312-13.

208

_____, "The Volcanoes of the Pacific Coast of the United States," in *Journal of the American Geographical Society, 1877, IX* (Albany, New York, 1879), 45-65.

(6) August 16-17, 1883. See pages 60-68.

Party: P. B. Van Trump, George B. Bayley, and James Longmire, guided to the mountain by Sotolick or "Henry," a Klickitat Indian.

Route: The mountain was approached by way of the Nisqually Valley to Paradise Park, where a base camp was established. An ascent, via the Gibraltar route, was made from a high camp below present Camp Muir. The climbers remained overnight in the small crater.

W. C. Ewing accompanied the party to the high camp but did not climb to the summit. A horse trail was cut to Paradise Park under Indian Henry's direction.

References: G. B. Bayley, "Ascent of Mount Tacoma," in *Overland Monthly*, VIII (September, 1886), 266-78.

P. B. Van Trump, "Mount Takhoma," in Tacoma *Every Sunday*, November 5, 12, 1892 (reprinted from Olympia *Transcript*, 1883).

(7) August 20, 1884. See pages 70-79.

Party: J. Warner Fobes, George James, and Richard O. Wells.

Route: The mountain was approached by way of the Bailey Willis trail to Spray Park where a base camp was established. Failure of an attempted ascent via Ptarmigan Ridge caused them to move their base camp to the divide between the Carbon and Winthrop glaciers, and from there, the summit was reached in a one-day ascent via the Winthrop-Emmons route.

The round trip required eleven hours.

References: J. Warner Fobes, "To the Summit of Mount Rainier," in *West Shore*, XI (September, 1885), 265-69 [reprinted by Theodore Gerrish in *Life in the World's Wonderland* (Biddeford, Maine, 1887), 185-208].

George James, "Mount Rainier," in Snohomish *Eye*, September 6, 1884 (reprinted in *Tacomian*, November 26, 1892).

(8) September, 1886. See pages 81-83.

Party: Allison L. Brown and six or seven Yakima Indians.

Route: The mountain was approached from the Yakima Reservation on Indian trails which led to Summerland. From there an ascent was made via Ingraham Glacier to an unknown elevation on the summit dome. An overnight camp was made at Cadaver Gap on the return.

This is an unverified ascent and there is nothing to indicate the summit was reached. The young, agency-educated Indians apparently climbed willingly enough.

Reference: Allison L. Brown, "Ascent of Mount Rainier by the Ingraham Glacier," in *The Mountaineer*, XIII (November, 1920), 49-50.

(9) August 11, 1888. See pages 84-94.

Party: P. B. Van Trump, John Muir, E. S. Ingraham, A. C. Warner, C. V. Piper, Daniel W. Bass, Norman O. Booth.

Route: The mountain was approached by way of the Yelm trail to Paradise Park where a base camp was established at Camp of the Clouds. The summit was reached in a one-day ascent via the Gibraltar route from a high camp at Camp Muir.

This party named Camp Muir, and the photographer, A. C. Warner, took the first photographs at Mount Rainier, including views of Longmire's Springs, Camp of the Clouds, and on the summit.

209

References: E. S. Ingraham, "Discovery of Camp Muir," in Meany's *Mount Rainier, A Record of Exploration* (New York, 1916), 150-58 (reprinted from *Puget Sound Magazine,* October, 1888).

_____, "Early Ascents of Mount Rainier," in *The Mountaineer,* II (November, 1909), 38-41.

John Muir, in *Picturesque California, the Rocky Mountains and the Pacific Slope,* Div. 6 (New York, c. 1888), 285-88 [reprinted in *Pacific Monthly,* VIII (November, 1902), 197-204, and in *Steep Trails,* ed. by W. F. Bade (Boston, 1918), 261-70].

C. V. Piper, "A Narrow Escape," in *The Mountaineer,* VIII (December, 1915), 52-53.

P. B. Van Trump, "On Top of Tacoma," in Tacoma *Ledger,* August 19, 1888.

_____, letter to George B. Bayley, dated August 26, 1888, in Oakland *Tribune,* July 23, 30; August 6, 30, 1939.

A. C. Warner, "A Climb to the Summit of Mount Rainier," six-page typescript at the John Muir School, Seattle, Wash. [published as "John Muir's Ascent of Mt. Rainier," in *The Mountaineer,* L (December, 1956), 38-45].

(10) August 13-14, 1889. See pages 97-102.

Party: E. S. Ingraham, E. C. Smith, Grant Vaughn, L. M. Lessey, Roger S. Green, H. E. Kelsey, and J. V. A. Smith.

Route: The mountain was approached by way of the Yelm trail to Paradise Park where a base camp was established. The summit climb was made via the Gibraltar route from a high camp at Camp of the Stars (Misery), and an overnight stay was made in the east crater.

This was the first party to make a high camp at Camp Misery and the first to stay overnight in the east crater.

References: E. S. Ingraham, "Mount Takhoma," in *Tacomian,* December 17, 1892 (reprinted from Seattle *Post-Intelligencer,* April 12, 1892).

E. C. Smith, "An Ascent of Mt. Rainier," in *Christian Register,* October 3, 1889, 4-5.

(11) August 23-24, 1889. See pages 102-105.

Party: Charles H. Gove and J. Nichols.

Route: The mountain was approached by way of the Yelm trail to Paradise Park where a base camp was established. The summit climb was made via the Gibraltar route from a high camp at Camp Muir, and an overnight stay was made in the east crater.

Gove was the first member of a mountain climbing organization to ascend Mount Rainier, and Nichols the first Tacomian. The tent they used at Camp Muir was the first "mountain tent" used on Mount Rainier.

References: C. H. Gove, "Night on the Summit," in Tacoma *Ledger,* September 1, 1889.

_____, "Night on the Summit of Mount Rainier," in W. G. Steel's *The Mountains of Oregon* (Portland, Oregon, 1890), 43-51 (reprinted in *Tacomian,* December 24, 1892).

(12) July 13, 1890. See pages 106-108.

Party: Oscar Brown.

Route: A one-day ascent, via the Gibraltar route, was supposedly made from Paradise Park.

This solo ascent is unverified and open to question.

References: Oscar Brown, "Mount Tahoma . . . Oscar Brown's Ascent in 1890," in *Tacomian*, March 25, 1893.

"Mount Tahoma . . . Historical Sketch of all Successful Climbers to the Above Date," in *Tacomian*, January 14, 1893.

(13) August 10-11, 1890. See pages 109-115.

Party: E. C. Smith, Fay Fuller, Leonard Longmire, W. O. Amsden, and R. R. Parrish.

Route: The mountain was approached by way of the Yelm trail to Paradise Park where a base camp was established. The summit climb was made via the Gibraltar route from a high camp at Camp Muir, with an overnight stay in the east crater.

Fay Fuller was the first woman to ascend Mount Rainier.

References: Fay Fuller, "A Trip to the Summit," in Tacoma *Every Sunday*, August 23, 1890 (reprinted in *Tacomian*, January 14, 21, 1893).

_____, (Mrs. Fritz von Briesen), Interview with Messrs. Bill, Potts, and McIntyre, of Mount Rainier National Park, August 17, 1950, at Tacoma, Washington.

Len Longmire, "An 'Old Timer' Recalls Some Interesting Happenings of the Past," in *Mount Rainier National Park Nature Notes*, XI (September, 1933), 4-5.

E. C. Smith, "A Trip to Mount Rainier," in *Appalachia*, VII (March, 1894), 185-205.

"She Reached the Top," in Tacoma *Ledger*, August 17, 1890.

"Mountain Climbers," in Tacoma *Ledger*, August 19, 1890.

"First Woman to Climb Mountain Visits City," in Tacoma *News-Tribune*, August 17, 1950.

(14) August 11-13, 1890. See pages 115-117.

Party: A. F. Knight, S. C. Hitchcock, Will Hitchcock, and F. S. Watson.

Route: The mountain was approached by way of the Yelm trail to Paradise Park where a base camp was established. The summit climb was made via the Gibraltar route, with a high camp near Camp Comfort, and an overnight stay in the east crater near Register Rock.

This party was the first to make a high camp above Gibraltar Rock and first to stay two nights on the summit dome.

References: A. F. Knight, Letter to C. F. Brockman, August 29, 1933; copy in Tacoma Public Library.

"Mount Tahoma . . . Two Iowa Editors Reach the Summit," in *Tacomian*, April 1, 1893 (reprinted from Tacoma *Ledger*, August 19, 1890).

Jim Whittaker, note to the author concerning inscriptions copied from rocks on the crater rim, August 13, 1954.

(15) July 2, 1891. See pages 120-122.

Party: Oscar Brown, A. G. Rogers, P. L. Markey, and Silas Balsely.

Route: The mountain was approached from Enumclaw by way of Lake Kapowsin and the Yelm trail to Paradise Park where a base camp was established. A one-day ascent to the summit was made from there.

This was the first party to ascend from Paradise Park without using a high camp. They carried a twenty-foot flagpole to the summit and left a flag and pennant flying there.

References: A. G. Rogers, "On the Way," in Enumclaw *Evergreen*, June 26, 1891.

_____, "Oscar Brown's Expedition," in Enumclaw *Evergreen,*
July 17, 1891.

_____, "A G. Rogers' Account of the 1891 Party," in *Tacomian,*
March 25, 1893.

_____, "First Flag on Mt. Tacoma," in Tacoma *Union,* June 9,
1895.

(16) July 29-31, 1891. See pages 122-123.

Party: Len Longmire, Elcaine Longmire, Susan Longmire, Edith Corbett,
E. T. Allen, E. A. Stafford, Hans Paulson, and Miller Cooper.

Route: Nothing is known of the approach routes of the several groups which
established base camps at Paradise Park and contributed members
to this ascent. The summit climb was made via the Gibraltar route
from a high camp in the vicinity of Camp Muir with an overnight
stay in the east crater.

On this ascent, Len Longmire began twenty years of guiding on
Mount Rainier.

References: E. T. Allen, Letter of Superintendent O. A. Tomlinson, of Mount
Rainier National Park, March 10, 1936.

Mrs. Susan Hall (*nee* Longmire), Interview with the author, May 24,
1955, at Tacoma, Washington.

E. S. Ingraham, *The Pacific Forest Reserve and Mt. Rainier,*
(Seattle, 1895) , 13.

"Special to the Ledger," in Tacoma *Ledger,* August 4, 1891.

"Those Mountaineers," in Tacoma *Ledger,* August 27, 1891.

Len Longmire, Interview with the author, September 28, 1954, at
Yelm, Washington.

(17) August 11-12, 1891. See pages 123-129.

Party: P. B. Van Trump, Warren Riley, and A. Drewry.

Route: The mountain was approached by way of the Yelm trail and Indian
Henry's trail to a base camp in the park known as Indian Henry's
Hunting Ground. The mountain was ascended via the Tahoma
Glacier, which was reached by crossing over from Success Cleaver. A
high camp was made below the cliffs at the point where they swung
onto the North Tahoma Glacier, and another night was spent in the
west crater before descending via the Gibraltar route.

This was the first ascent of Mount Rainier from the west side and
the first using a different route for the descent. Dr. Riley's deerhound
accompanied the climbers throughout.

References: "On the Mountain Top," in Tacoma *Ledger,* August 26, 1891.

"Those Mountaineers," in Tacoma *Ledger,* August 27, 1891.

"Climbing Mount Rainier," in Tacoma *Washington Standard,*
September 18, 1891 (reprinted from the Centralia *News*) .

P. B. Van Trump, "On the Mountain Top," in Tacoma *Ledger,*
August 31, 1891 (reprinted in Tacomian, February 25, 1893).

(18) August 20, 1891. See pages 129-132.

Party: Frank Taggart, Grant Lowe, and Frank Lowe.

Route: The mountain was approached from Orting by way of Lake Kapowsin,
the Yelm trail, and Indian Henry's trail, and a path they cut through
the forest to the terminus of the Tahoma Glacier where they en-

camped. From there, they climbed the glacier to the summit, making an overnight stop on the ascent but none on top.

Riley and Drewry met this party on the Puyallup River on the 17th.

References: Frank Lowe, "Mount Tahoma; an Orting Party Reached the Summit in August 1891 via the Succotash Valley," in *Tacomian*, March 18, 1893 (reprinted from Orting *Oracle*, September, 1891).

(19) July 29, 1892. See pages 134-136.

Party: George L. Dickson, W. H. Dickson, H. W. Baker, W. G. Cassels, and W. E. Daniel.

Route: The mountain was approached from Tacoma by the county road and the Succotash Valley trail to Paradise Park, where a base camp was established. The summit climb was made via the Gibraltar route from a high camp at Camp Misery, with a return the same day.

This may have been the first climbing party to travel directly to the mountain from Tacoma. They gave the campsite at foot of Gibraltar Rock its name, "Misery."

References: George L. Dickson, "Mount Tahoma . . . George L. Dickson's Narrative," in *Tacomian,* January 7, 1893.

"Climbing Mt. Tacoma," in Tacoma *Ledger,* August 3, 1892.

"Mountain Murmurs," in Tacoma *Every Sunday,* August 6, 1892.

(20) July 30-31, 1892. See pages 136-138.

Party: Warren Riley and George Jones.

Route: The mountain was approached by way of the Yelm trail and Indian Henry's trail to a base camp in Indian Henry's Hunting Ground. The summit climb was made up the Tahoma Glacier, with an overnight stop on the ascent and a second night in the west crater, which they shared with the Taggart-Lowe party. The two parties joined in the conquest of the North Peak on the following day.

This was the first time two parties were together on the summit. They were also first to reach Liberty Cap (North Peak).

References: "Mountain Murmurs," in Tacoma *Every Sunday,* August 13, 20, 1892.

(21) July 30-31, 1892. See pages 137-138.

Party: Frank Taggart and Frank Lowe.

Route: Their approach to the mountain, and summit climb probably followed the route of 1891. It is also likely that they joined the Riley-Jones party on the Tahoma Glacier, continuing with them.

These climbers removed, as a souvenir, the lead plate left in the west crater in 1883.

References: See the references given for the Riley-Jones ascent.

P. B. Van Trump, "Conquering the Giant," in Tacoma *Ledger,* September 12, 1892.

(22) August 20, 1892. See pages 140-141.

Party: James G. Van Marter and George B. Hayes.

Route: Nothing is known of the approach route, though their base camp was in Paradise Park. The first attempt to reach the summit was foiled by a storm which caught them at Camp Misery, but they were successful on the second try—a one-day ascent via the Gibraltar route from a high camp at Camp Misery.

213

Van Marter was the first climber with European experience to reach the summit, and he introduced the use of sleeping bags on the mountain.

References: "Mountain Murmurs," in Tacoma *Every Sunday*, September 3, 1892.

"All night in a Storm," in Tacoma *Ledger*, August 28, 1892.

(23) August 21-22, 1892. See pages 141-145.

Party: P. B. Van Trump and George B. Bayley.

Route: The route of the previous year was retraced by Van Trump in the approach to the mountain, but the base camp was located higher— near timber line on the Success Cleaver. On the summit climb, the crossing from the cleaver to Tahoma Glacier was made at a lower elevation than on Van Trump's earlier ascent; otherwise the route followed was much the same. One overnight camp was made on the ascent and another night was spent in the west crater, after which the North Peak was visited before descending. Bayley was seriously injured by a fall into a crevasse while descending the Tahoma Glacier and his survival is one of the epic stories of those early days.

References: "Mountain Murmurs," in the Tacoma *Every Sunday*, July 23, August 13, 1892.

"Like a Flash an Oaklander's Awful Slide Down Mt. Rainier," in Oakland *Enquirer*, September 23, 1892.

P. B. Van Trump, "Conquering the Giant," in Tacoma *Ledger*, September 12, 1892 (reprinted in *Tacomian*, March 11, 1893).

————, "Mount Tahoma," in *Sierra Club Bulletin*, I (May, 1894), 109-32.

(24) August, 1892. See page 138.

Party: W. W. Seymour and "guide" Carter.

Route: Gibraltar route.

Very little is known about this ascent.

Reference: "Home from the mountain," in Tacoma *Ledger*, September 1, 1892.

(25) September 3-4, 1892. See page 146.

Party: Len Longmire and Lawrence Corbett.

Route: The ascent was made via the Gibraltar route with an overnight stay in the east crater.

Snow fell during the night and covered some of their equipment, which was lost.

References: Susan Hall (*nee* Longmire), Interview with the author, May 24, 1955, at Tacoma, Washington.

Len Longmire, Interview with the author, September 28, 1954, at Yelm, Washington.

(26) August 29-30, 1893. See pages 165-167.

Party: W. M. Bosworth, Arthur French, Guy Evant, and Walter Wolff.

Route: The ascent was made from Camp of the Clouds, via the Gibraltar route, without the use of a high camp. A night was spent in the west crater.

This party found the Oscar Brown flag and reset it on the summit.

214

References: W. M. Bosworth, "Slept in the Crater," in Tacoma *Ledger,*
October 2, 1893.

"Along Paradise River," in Tacoma *Ledger,* September 4, 1893.

"Planted on Top Mt. Tacoma," in Tacoma *Union,* May 26, 1895.

(27) July 18-19, 1894. See pages 175-178.

Party: E. S. Ingraham, Leo Daft, H. P. Strickland, F. W. Hawkins, Annie
Hall, M. Bernice Parke, Helen Holmes, H. E. Holmes, W. H. Wright,
G. E. Wright, N. A. Carle, L. M. Lessey, Roger Greene, Jr., and Ira
Bronson.

Route: From its base camp at Camp of the Clouds, this party ascended the
mountain via the Gibraltar route, using Camp Muir as a high camp.
An overnight stay was made in the east crater.

References: E. S. Ingraham, "From the Ice Caves," in Seattle *Post-Intelligencer,*
July 25, 1894.

_____, "It Rises Above All," in Seattle *Post-Intelligencer,*
August 12, 1894.

_____, *The Pacific Forest Reserve and Mt. Rainier* (Seattle,
1895) , 19-20.

(28) August 8, 1894. See pages 179-181.

Party: Olin D. Wheeler, George M. Weister, Lyman B. Sperry, Henry M.
Sarvent, and Ross Comstock.

Route: The base camp of this party was on Theosophy Ridge, and the
summit climb was made via the Gibraltar route with a high camp
near the Beehive. They did not stay overnight on the summit.

This ascent was sponsored by the Northern Pacific Railroad in order
to encourage tourist travel.

Reference: O. D. Wheeler, "Mount Rainier. Its Ascent By a Northern Pacific
Party," in *Wonderland* (1895) , 52-103.

215

NOTES

1 "Journal of Occurrences at Nisqually House, 1833," *Washington Historical Quarterly*, VI (July, 1915), 186.

2 His diary entry for June 23, 1833, is an example: "Had a splendid view of Mt. Rainier, whose gold vestures it was most refreshing to behold from the sultry hollows of prairie."

3 W. F. Tolmie, Ms. Diary of Dr. William Fraser Tolmie, 1833; Photostats from original pages 11-21, the Provincial Archives, Victoria, B. C.

4 Tolmie diary, 11.

5 According to Tolmie's diary entry for August 29, the encampment was on the "prairie 8 miles N. of House," which would put it near the north end of Steilacoom Lake. At that point, the old Indian trail, which paralleled Puget Sound from the Nisqually River crossing, turned westward toward Muckleshoot Prairie on White River; hence, it is likely they followed that primitive trace as far as the Puyallup River crossing near the present town of McMillan. Moderation of the climate since that time has assisted in the forestation of the once relatively open Nisqually plain.

6 The horses were probably left at the southern end of the alluvial bottom which terminates several miles south of Orting, near where the Puyallup River debouches from a narrow canyon.

7 Tolmie's description would indicate the campsite was near the site of the present bridge by which the Orting-Kapowsin highway crosses the Puyallup River, for it is the only constriction along that part of the stream.

8 The three-mile-long canyon above present Electron was an obstacle more easily detoured than passed through.

9 Probably the prominent hills in sections 8 and 19, Township 17 North, Range 6 East.

10 A capote was a hooded coat or cloak.

11 On the Mowich River, approximately two miles west of the present boundary of Mount Rainier National Park. They had made eight miles during the day.

12 Paul Peak, the Hessong Rock-Mount Pleasant summit, and the unnamed ridge between the North and South Mowich rivers. The peak shown on present maps as Tolmie Peak cannot be seen from any point along the Mowich River above the National Forest boundary; hence it was not one of the "snowy peaks above" referred to by Tolmie.

13 The conformation of the hills surrounding the forks of the Mowich River gives a definite amphitheatre effect to that point. In checking that portion of Tolmie's route within the national park, I had the good fortune to make one trip over it under almost identical weather conditions. Above the forks, several waterfalls were visible plunging from the south wall of the canyon, but the really spectacular one thundered over a three hundred-foot drop on the medial line of

the North Mowich Glacier, just below the 7,000 foot elevation. Several days of heavy fall rain had turned the runoff from the lower reaches of the glacier into a mighty, brownish cataract. Tolmie's view of the river "bounding over a lofty precipice above" must have been the same sight that I watched with Gene Faure, of the Tacoma Mountaineers, on that memorable scramble in the fall of 1955.

14 I believe this encampment was at the head of the drainage between Fay Peak and Hessong Rock; lack of any mention of a lake is strong argument against the Eunice Lake site, and the Hessong Rock-Mount Pleasant summit agrees with Tolmie's description as well as Tolmie Peak does.

15 Lachalet was always a staunch friend of Tolmie and he remained a man of importance with the Hudson's Bay and Puget Sound Agricultural companies.

16 Illegible, but probably "dead."

17 Crevasses.

18 Echo and Observation rocks.

19 September 5, 1833; *WHQ*, VI:195.

20 Charles Wilkes, U.S.N., *The United States Exploring Expedition*, IV (Philadelphia, 1845), 309, 414.

21 Wilkes, *Exploring Expedition*, IV:413.

22 On the basis of the accepted elevation of 14,410 feet, the difference is 2,080 feet, of which approximately 1,160 feet can be attributed to omission of the first two corrections, while an error of about 200 feet would result from failure to use the true elevation of the instrument station.

23 Olympia *Columbian*, September 18, 1852.

24 Hubert Howe Bancroft, *History of Washington, Idaho, and Montana,* 1845-1889 (San Francisco, 1890), 386.

25 Hazard Stevens made a similar estimate of ten miles for his climb with Van Trump in 1870. See "Mountain Notes," in Tacoma *Every Sunday*, August 23, 1890.

26 Reflection Lake, Lake Louise, and Bench Lake, if Bancroft is correct.

27 "What-a-Man Named Bailey," in Seattle *Times*, July 7, 1935.

28 Leslie M. Scott, "News and Comments," in *Oregon Historical Quarterly*, XIX (December, 1918), 337-38.
George H. Himes, "Very Early Ascents," in *Steel Points*, I (July, 1907), 199.

29 Shaw was born in Missouri in 1829, and emigrated to Oregon in 1844.

30 Olympia *Columbian*, April 23, May 28, and June 18, 1853.

31 Winthrop wrote, "Of all the peaks from California to Frazer's River, this one before me was royalest. Mount Regnier Christians have dubbed it, in stupid nomenclature, perpetuating the name of somebody or nobody. More melodiously the Siwashes call it Tacoma—a generic term also applied to all snow peaks." (1913 ed., p. 36.)
In 1868, General M. M. McCarver gave the name of Tacoma to the townsite he platted on Commencement Bay, but the attempt to apply the name to Mount Rainier began with an announcement in the March, 1883, *Northwest Magazine*, that "The Indian name Tacoma will hereafter be used in the guide books and other publications of the Northern Pacific Railroad Company." See Conover's "Mount Rainier and the Facts of History," p. 19.)
Winthrop's name has been given to the great glacier which feeds the West Fork of White River; formerly it and the Emmons Glacier were collectively known as the "White River glaciers."

[32] Clinton A. Snowden, *History of Washington* (New York, 1909), III : 162.

[33] "Narrative of James Longmire: A Pioneer of 1853," in *Washington Historical Quarterly*, XXIII (January, April, 1932), 47-60, 138-60. This account has been altered slightly from the original published as a part of the "Old Settler Stories," in Tacoma *Ledger*, August 21, 1892 (from Mrs. Lou Palmer's notes of a personal interview with James Longmire).

[34] Kate L. Gregg, "The Saga of Lolo Stik: Part I, William Packwood—Pioneer," in Seattle *Times*, October 14, 1951.

[35] Frances Kautz, "Diary of Gen. A. V. Kautz," in *Washington Historian*, I (April, 1900), 115.

[36] Joseph McEvoy, "An Old Soldier's Story," in Tacoma *Ledger*, May 28, 1893.

[37] A. J. Splawn, *Ka-mi-akin, The Last Hero of the Yakimas* (Portland, c. 1917), 340, has recorded it as "one or two years before the Indian War of 1855," while Lucullus V. McWhorter, "Chief Sluiskin's True Narrative," in *Washington Historical Quarterly*, VIII (January, 1917), 96-101, has it one or two years after the Walla Walla Council of May and June, 1855, but that is impossible. It could only have been 1855, if it *was* after the council, and I prefer that date as it agrees with the better account, carefully taken through two interpreters.

[38] Splawn estimates Saluskin's age as "about twenty" at the time of this story, which would place his birth between 1833 and 1836; however, Yakima Indian Agency records give it as 1823.

[39] The Walla Walla Council, at which the young man Saluskin cared for the horses of Chief Owhi.

[40] He means the site of the present Moxee Bridge, about two miles east of North Yakima.

[41] Fielding Mortimer Thorp, first permanent settler in the Yakima Valley, where he located February 15, 1861.

[42] Tal-e-kish.

[43] Splawn says, "the spot which is now the fine ranch of John Russell in the Tieton basin."

[44] The terminus of the Emmons Glacier.

[45] Splawn says they also killed a kid just before going into camp. When Saluskin asked why they didn't kill one of the larger goats, he was told the younger ones were better to eat.

[46] Yakima Park.

[47] They were at Mystic Lake.

[48] An avalanche at the top of Willis Wall.

[49] Splawn says that upon returning the two white men ate only a few bits of bread, then lay down and slept. They remained in camp all the next day, did some writing and drew a map on a large paper, asking Saluskin the names of the different streams flowing into the Yakima River, which he told them.

[50] Apparently a reference to the story he told to A. J. Splawn.

[51] A number of newspaper accounts concerning the chief appeared in September, 1915; they were grossly inaccurate.

[52] James Longmire, "James Longmire, Pioneer," in Tacoma *Ledger*, August 21, 1892.

[53] A. J. Frost, W. D. Vaughn, and William Martin, "Old Settler's Stories," in Tacoma *Ledger*, July 22, 1892, February 12, 1893, and March 26, 1893.

[54] Hazard Stevens, *Life of General Isaac I. Stevens* (Boston, 1901), II : 201.

55 A. V. Kautz, "The Death of Leschi," in Tacoma *Ledger,* April 9, 1893.

56 A. V. Kautz, "Ascent of Mount Rainier," in *Overland Monthly,* XIV (May, 1876), 393-403. This account is used frequently hereafter without citation.

57 A. V. Kautz, Ms. Journal of Lt. August V. Kautz, 1857-61; Microfilm A-383, 8-12, the National Archives. Referred to hereafter without citation.

58 Wren was one of the former Hudson's Bay Company employees suspected of aiding the Indians during the war of 1855-56; his homestead was east of Roy in Section 30, T18N, R3E, Willamette Meridian.

59 The statement Kautz makes in his "Ascent of Mount Rainier," to the effect that it had not rained for weeks is not substantiated by his diary.

60 Each cracker was about five by five inches and one-half inch thick, of unleavened bread, looking much like a large soda wafer.

61 The encampment was between Elbe and Mineral Junction.

62 The terminus of the Nisqually Glacier was then about 300 feet below the present highway bridge (1956). A careful search of the Kautz papers in the National Archives failed to locate the sketch he mentions.

63 This observation is not recorded in Kautz's diary. It gives an elevation which is approximately 1,400 feet too high, probably due to the low barometric pressure of the developing storm.

64 George Gibbs, "Physical Geography of the Northwestern Boundary of the United States," in *Bulletin of the American Geographical Society of New York* (1872), 356-57. The statement is attributed to Kautz.

65 A. V. Kautz, "Expedition to the Summit of Mt. Ranier," in Steilacoom *Washington Republican,* July 24, 1857.

66 Gibbs, "Physical Geography," *loc. cit.*

67 Kautz states that he reached an elevation of 12,000 feet, but that was based on his belief that the height of Mount Rainier was 12,330 feet, as determined by Lieutenant Charles Wilkes in 1842. "He Reached the Top," in Tacoma *Ledger,* January 31, 1893, for a clarification of this much-debated point.

68 John S. Hittell, "Some Notes on the Resources, Population and Industry of Washington Territory," in *The Hesperian,* III (1859), 398.

CHAPTER II

pp. 29-57

1 Clinton A. Snowden, *History of Washington,* IV:193.

2 Wm. Packwood, "Important Territorial Road," in Steilacoom, W. T., *Puget Sound Herald,* July 15, 1859.

3 *Ibid.* "Tuesday, June 21st [1859]—Left camp on a stream flowing eastward into the valley of the Nachess, and about five miles above where the same opens out into quite a large valley; travelled up the same in a W.N.W. direction ten miles, and camped at the head of the stream where it takes its rise in Bear Prairie. At this point the valley is perhaps half a mile wide . . . Bear Prairie is situated immediately on the summit of the Cascade mountains on this route."
"*Wednesday,* June 22d—Left camp; travelled same direction; passed Bear Prairie; then down an alder and skunk cabbage muck valley about thirty rods wide, of slight descent, some two miles and came to the upper Nisqually valley, which lies on the Nisqually River below where the same leaves the mountains south and east of Mount Rainier."

4 Biographical data were obtained from Hazard Stevens' *Life of General Isaac I. Stevens*, pp. 81, 438 and 456; and from Francis B. Heitman's *Historical Register and Dictionary of the United States Army* (Washington, D. C., 1903), I:922. The United States Department of the Army's publication *Medal of Honor* (Washington, D. C., 1948) , p. 123, has the following entry concerning Hazard Stevens' award: "Rank and Organization: Captain and Assistant Adjutant General, U. S. Volunteers. Place and Date: At Fort Huger, Va., 19 Apr. 1863. Entered Service at Olympia, Washington Territory. Birth: Newport, R. I. Date of Issue: 13 June 1894. Citation: Gallantly led a party that assaulted and captured the fort."

5 Biographical data from "Death Comes to P. B. Van Trump," in Tacoma *News Tribune*, December 28, 1916.

6 P. B. Van Trump, "Mount Rainier," in *Mazama*, II (October, 1900), 4.

7 Biographical data from George H. Himes, "Discovery of Pacific Coast Glaciers," in *Steel Points*, I (July, 1907) 146-48; and A. L. Mumm (compiler) , *The Alpine Club Register*, 1857-63 (London, 1923) .

8 Hazard Stevens, "The Ascent of Takhoma," in *Atlantic Monthly*, XXXVIII (November, 1876) , 511-30. This account is used frequently hereafter without citation.

9 Van Trump, "Mount Rainier," see footnote 6. This account is used frequently hereafter without citation.

10 Laura Virginia Wagner, *Through Historic Years with Eliza Ferry Leary* (Seattle, 1934) , 14.

11 The flag is on exhibition at the Washington State Historical Society's museum at Tacoma.

12 "On Mount Tacoma—one of the flags placed there in 1870 owned by W. H. Cushman," in Tacoma *Ledger*, February 17, 1891.

13 Both Van Trump and Coleman say he rode a horse, but later details support Stevens.

14 That attack on the small band of Nisqually Indians led by Ski-hi is one of the least known events of the Indian war, and consequently, one of the most misrepresented. It was not a "massacre" but a surprise attack of a sort which was a standard tactic of Indian warfare and used by both white and red men.

15 The "old Indian trail" previously referred to. In earlier times it was one of the arteries for traffic between the Indians of Puget Sound and those of eastern Washington.

16 Such a failure was often construed as malpractice. Poniah, headman of a little band of Upper Cowlitz Indians is said to have died for a similar reason. An aggrieved relative of a dead patient shot him, but found the old man too slow dying; so he took an arrow and churned it up and down in Poniah's throat to stop the "cultus wahwah," or bad talk with the spirits.

17 The stream cannot be identified now.

18 George Gibbs described the upper Cowlitz Indians as a hermit-like branch of the Klickitat tribe, about seventy-five in number, who "avoided all intercourse with their own race." According to him, their less wild relatives attached to them "all kinds of superstitious ideas, including that of stealing and eating children, and of traveling unseen." "Report to Capt. George B. McClellan, at Olympia, W. T., March 4, 1854," cited by R. N. McIntyre, in *Short History of Mount Rainier* (Longmire, Washington, 1952) , 8.

19 The point marked with the elevation 4751, on the *Mount Rainier Quadrangle*, N4630-W12130/30, edition of 1928, published by the U.S.G.S.

20 Now known as Tatoosh Creek; it is the outlet of Reflection Lake.

21 Stevens and Van Trump do not agree on which encampment they were at when Sluiskin made his harangue, but I have followed Stevens who wrote only six years after the event, and whose narrative is generally more detailed and exact.

22 Congressman M. C. George of Oregon committed the Chinook to memory soon after the return of the successful climbers. A copy later furnished Stevens became the basis of a translation in 1915.

23 Now the Paradise River.

24 Present day climbers are familiar with the great summit crevasse which extends in a sweeping arc from near Point Success across the heads of the Kautz, Nisqually, Ingraham, Emmons, and Winthrop glaciers. It has stopped many less resourceful parties.

25 The plate and canteen were not found on subsequent ascents, though Van Trump and Stevens both searched for them. Perhaps the ice mantle of the summit was thinner, exposing more of the rocky crown, which would explain why the west crater was so much more prominent then than now.

26 The ascent from there to Peak Success had required seven and one half hours.

27 Herbert Hunt, *Tacoma, Its History and Its Builders* (Chicago, 1916), I:162-63.

28 There is complete confusion about this name: it appears as Captain Turrell, Captain F. W. Ferrell, and Captain Terrill, but he probably was Freeman W. Tyrrell of Hawk's Prairie.

29 "Mountain Notes," in Tacoma *Every Sunday*, August 23, 1890.

30 Mrs. George E. Blankenship, *Early History of Thurston County, Washington* (Olympia, Washington, 1914) , 389. ". . . Captain F. W. Ferrell claimed to have ascended it in October 1849, in company with John Edgar, and a Frenchman and Indian, names unknown." Interesting, since he didn't come to the Puget Sound country until 1851.

31 The 1856 observations were made from a station in the vicinity of Port Townsend, employing a line of sight 98.5 miles long. The approximate elevation computed for the middle summit was 14,359 feet. (From the records of the United States Coast and Geodetic Survey.)

32 S. F. Emmons, "The Volcanoes of the Pacific Coast of the United States," in *Journal of the American Geographical Society*, 1877 (Albany, New York, 1879) , 45-65. Used hereafter without citation.

33 Emmons says the face of the ice was then 500 feet in height.

34 A meaningless statement in view of the superficial use made of climbing ropes by all except a few expert climbers in the Alps.

35 Wilson was only temporarily defeated. While in the employment of the Northern Transcontinental Survey many years later, he computed the elevation of Mount Rainier as 14,900 feet, on the basis of 100 angulations from the east. (See Ingraham's "It Rises Above All," in Seattle *Post Intelligencer*, August 12, 1894.)

36 He did not terminate his work in California in time to follow them, as originally planned.

37 S. F. Emmons, "Glaciers of Mt. Rainier," in *American Journal of Sciences and Art, CI* (March 1871), 157-67.

38 S. F. Emmons, "Ascent of Mt. Rainier, with Sketch of Summit," in *The Nation*, No. 595 (November 23, 1876) , 312-13.

39 See footnote 31.

pp. 58-79

1 His map was printed with a geological report in the Tenth Census of the United States (1880), and is reproduced in the Washington, D. C., *Evening Star,* January 13, 1894.

2 According to R. N. McIntyre, a party of eastern Washington cattlemen, and their friends, from what is now known as Union Gap, made a leisurely trip into the parks at the foot of Little Tahoma in August of 1881. Under the leadership of P. J. Flint, and guided by a Yakima Indian, the group travelled up the Tieton River and Indian Creek to Cowlitz Pass, then followed down Summit Creek and the Ohanapecosh River to the point of Backbone Ridge, up which they passed to the Cowlitz Divide and Cowlitz Park. Horses were taken across the Whitman Glacier to an encampment on some bare rocks overlooking the Ingraham Glacier, but no attempt was made to climb Mount Rainier. [A. L. Flint, letter to Julius Halverson, August 26, 1938, cited in *Short History of Mount Rainier* (Longmire, Washington, 1952), 21.]

3 Quoted by C. T. Conover, "Mount Rainier and the Facts of History," in *The Great Myth—'Mount Tacoma'* (Olympia, Washington, 1924), 19.

4 George B. Bayley, "Ascent of Mount Tacoma," in *Overland Monthly,* VIII (Sec. Ser., September, 1886), 266. This account is used frequently hereafter without citation.

5 John Muir, "The Kings River Valley," in *Sierra Club Bulletin,* XXVI (February, 1941), 4 (reprinted from the San Francisco *Daily Evening Bulletin,* August 13, 1875).

6 Richard M. Leonard and David R. Brower, "A Climber's Guide to the High Sierra," in *Sierra Club Bulletin,* XXV (February, 1940), 43.

7 "Save Paradise Valley," in Tacoma *Ledger,* September 2, 1892.

8 Pacific Incubator Company. He is listed as a "capitalist" from 1888 to 1891.

9 P. B. Van Trump, "Mount Takhoma," in Tacoma *Every Sunday,* November 5, 12, 1892 (reprinted from Olympia *Transcript,* 1883). This account is used frequently hereafter without citation.

10 Nels Bjarke, "The Indian Henry Trail," mimeographed by the Fern Hill Historical Society (Tacoma, 1940), p. 6.
The origin of Indian Henry's "white" name is given in Herbert Hunt's *History of Tacoma* (p. 337), from which it appears that James Longmire, James Packwood and Henry Windsor were returning from a trip over the Cascade Mountains during the sixties when they met an Indian of the upper Cowlitz band of Klickitats on Skate Creek, south of Mount Rainier. Upon asking his name, they were told it was "Sotulick." Windsor then asked his "Boston name" and was told he didn't have one, whereupon Windsor gave him one—Henry—because he thought the Indian name too difficult.

11 Fay Fuller, "A Trip to the Summit," in Tacoma *Every Sunday,* August 23, 1890.

12 Chinook jargon meaning, "Soldiers! Go swiftly!" Those spiteful little creatures were warlike enough to be called soldiers without wearing the yellow stripes of the cavalry.

13 From "So-ho-tash," the place of the wild raspberry. That part of the Nisqually Valley was later known to the settlers as the Succotash Valley, from a corruption of the Indian name.

14 The name is still in use. Van Trump also mentions Goat Creek.

15 Where Longmire, the headquarters of Mount Rainier National Park is now located. The statement that the springs were found on the return trip can be traced to an article by Joseph Stampfler, "Climbing Mount Tacoma," in Tacoma News, July 12, 1907. Since Stampfler's version is only hearsay, I have followed Bayley.

16 The horses were taken to the edge of the Muir snowfield below Anvil Rock.

17 Kloshe Nannitch, the Chinook jargon equivalent of "be careful."

18 The rocky point east of Camp Muir. They camped in the slide rock at its base at an elevation of 9,800 feet (not at 11,300 feet as Bayley says).

19 Van Trump carried a hatchet, a six-foot flagstaff cut from an alpine fir, and a flag wrapped around his waist; Bayley had 100 feet of new manila rope; Longmire carried the whisky flask; and Ewing brought up the rear with the barometer.

20 Cowlitz Cleaver.

21 Gibraltar Rock, named by Major Ingraham because of its resemblance to the guardian of the Mediterranean Sea.

22 Near the point known as "register rock," elevation 14,161 feet.

23 In his article published in the Overland Monthly in 1886, Bayley gives the date as August 16, 1884, an error which has given rise to much confusion.

24 Recorded by George James who saw the inscription on August 20, 1884, when he climbed to the summit with J. Warner Fobes and Richard O. Wells. See the James account which follows.

25 Personal belongings.

26 Chinook jargon for a "happy heart."

27 Bailey Willis, A Yanqui in Patagonia (Stanford, California, c. 1947), 23-24.

28 Edmond S. Meany, ed., Mount Rainier, A Record of Exploration (New York, 1916), 290-91.

29 Willis, A Yanqui in Patagonia, 25.

30 Snohomish, W. T., Eye, September 12, 1883.

31 William Whitfield, History of Snohomish County, Washington (Chicago, 1926), I:255.

32 Snohomish, W. T., Eye, July 25, 1885; August 28, September 4, and September 18, 1886.

33 Aubrey L. Haines, "Three Snohomish Gentlemen Make a Successful Ascent," in The Mountaineers, XLVII (December, 1954), 6.

34 J. Warner Fobes, "To the Summit of Mount Rainier," in The West Shore, XI (September, 1885), 265. This account is used frequently hereafter without citation.

35 George James, "Mount Rainier," in Snohomish, W. T., Eye, September 6, 1884. This account is used frequently hereafter without citation.

36 In his article, "The Mount Blanc of our Switzerland," in Outing (February, 1885), 323, J. R. W. Hitchcock mentions "the four houses, forming the town of Wilkeson." He also speaks with kindly humor of a Mr. Jones, who was station agent, storekeeper, innkeeper, postmaster, express agent, and justice of the peace.

37 A camp for trail laborers with huts made of that material. The log cabin and barn later built there became known as the "Grindstone Camp."

223

38 Both Fobes and James use that name for the body of water now called Mowich Lake, and Fobes says there were fish in it then.

39 Fay Peak, named for Fay Fuller, the first woman to reach the summit of Mount Rainier; it appeared on the Sarvent map of 1895.

40 Spray Falls.

41 The great cirque of Carbon Glacier.

42 Sketches made by Bailey Willis in 1882 show that the terminus of the glacier was below the mouth of the gorge between the Northern Crags and Echo Cliffs. (Original sketches in the library of Mount Rainier National Park.)

43 At that time the Winthrop and Emmons glaciers were collectively termed the "White River glaciers" because they both nourished forks of that stream.

44 Probably due to the freezing of moisture condensed on the delicate system of chains and levers used to translate changes in atmospheric pressure into readings on a calibrated dial.

45 Cf., p. 94.

46 Van Trump examined the lead sheet again in 1891 and "found it whitened and corroded by its sulphur steam bath of eight years. On the reverse side some successful climber had since inscribed his name, W. Fobes." (See P. B. Van Trump, "The North Peak—Van Trump's First Attempt to Reach it, in August, 1891," in *Tacomian*, February 25, 1893.)

47 They had two pairs of goggles but did not use them, depending entirely on charcoal rubbed on the skin around the eyes.

48 From Mrs. Maude Shaffer (*nee* Longmire) interviewed by the author at Tacoma, Washington, May 3, 1955.

CHAPTER IV

pp. 80-96

1 Mrs. Jameson was a semi-invalid and spent several summers at Soda Springs. She is the woman who is seated in the rocking chair in the foreground of the photograph taken by A. C. Warner in 1888.

2 Edmond S. Meany, *Mount Rainier, A Record of Exploration* (New York, 1916), 302, *et seq.*

3 *Ibid.*

4 Allison L. Brown, "Ascent of Mount Rainier by the Ingraham Glacier," in *The Mountaineer*, XIII (November, 1920), 49-50.

5 Probably "Cadaver Gap."

6 "Great Glacier is a Monument to Seattle Man," in Seattle *Times*, July 19, 1925.

7 See accompanying map, *Gletscher des Mount Tacoma, nach dem Northern Transcontinental Survey*, von Bailey Willis.

8 E. S. Ingraham, *The Pacific Forest Reserve and Mt. Rainier* (Seattle, 1895) , 12. There were eleven men and eleven women in the party, and the women were Mrs. F. E. Nichols, Misses Emma, Carrie, and Mamie Shumway, Miss Edith Sanderson, Miss Minta Foster, and Miss Lena Rhine. See C. T. Conover, "Just Cogitating," in Seattle *Times*, July 29, 1956.

9 "First Woman to Climb Mountain Visits City," in Tacoma *News-Tribune*, August 17, 1950.

10 Interview with Mrs. Fritz von Brieson (*nee* E. Fay Fuller), August 17, 1950 (transcript in the files of Mount Rainier National Park).

11 Fay Fuller, "A Trip to the Summit," in Tacoma *Every Sunday*, August 23, 1890.

12 William G. Steel, *The Mountains of Oregon* (Portland, 1890), 67; also, "Alpine Club organized," in Portland *Oregonian*, September 15, 1887.

13 Linnie Marsh Wolfe, ed., *John of the Mountain* (Boston, 1938), 282.

14 Brother Cornelius, *Keith, Old Master of California* (New York, c. 1942), 167.

15 Aubrey L. Haines, ed., "John Muir's Ascent of Mt. Rainier, as Recorded by his Photographer, A. C. Warner," in *The Mountaineer*, L (December, 1956), 38-45. This account is used frequently hereafter without citation.

16 P. B. Van Trump, letter to George B. Bayley, August 26, 1888 (pub. in Oakland *Tribune*, July 23 to August 13, 1939). This account is used frequently hereafter without citation.

17 John Muir, ed., *Picturesque California, the Rocky Mountains and the Pacific Slope*, Div. 6 (New York, 1888), 286-88. This account is used frequently hereafter without citation.

18 John Hays helped build the "homestead cabin" at Longmire Springs in 1889. See "Narrative of James Longmire: A Pioneer of 1853," in *WHQ*, XXIII:47-56, 138-50.

19 Fourteen years old according to E. S. Ingraham's "Then and Now," in *The Mountaineer*, VIII (December, 1915), 50-51.

20 Brother Cornelius, *loc. cit.*

21 *Tacomian*, Jan. 21, 1893; "Personal. Allen C. Mason, Frank Ross, and F. S. Harmon, climbed the mountain to a height of 13,050 feet, during the week ending June 22, 1888, with James Longmire as guide."

22 Joseph Stampfler, "Climbing Mount Tacoma," in Tacoma *News*, July 12, 1907.

23 The original Camp of the Clouds was at an elevation of 5,700 feet and should not be confused with the tent camp of the same name established later on the south side of Alta Vista (elevation 5,530 feet).

24 He had a large view camera which was crude by present standards. It was eighteen inches long, with a body twelve inches square; the lens had a speed of f.22 and lacked a shutter (exposures were made by plucking a black cap from the lens and replacing it after the exposure was made). The camera with its tripod, glass plates and other items, weighed over fifty pounds.

25 Muir was carrying his favorite aneroid barometer, which gave him a reading of 10,000 feet.

26 E. S. Ingraham, "Discovery of Camp Muir," in Meany's *Mount Rainier*, 150-58. It was reprinted from the *Puget Sound Magazine*, edited jointly by Edmond S. Meany and Alexander Beggs and published in Seattle. The magazine probably perished with the Seattle fire, for no file of it is known to exist, though the Faculty Records of the University of Washington refer to it.

27 In his brief article, "On Top of Tacoma," in Tacoma *Ledger*, August 19, 1888, Van Trump gives the impression Warner brought up the rear, but other accounts establish that it was Piper.

28 Piper graduated from the University of Washington in 1885 and became a professor of botany and zoology at Washington Agricultural College, 1892-1903. His version of the near-accident was given in "A Narrow Escape," in *The Mountaineer*, VIII (December, 1915), 52-53.

29 Ingraham indicates that Norman Booth died prior to February 1893. See "Mr. Ingraham's Letter," in *Tacomian*, February 4, 1893.

30 The measurement was made by James Smyth Lawson, member of a Coast Survey party under Prof. G. C. Davidson, from a base line near Puget Sound. See E. S. Ingraham, "It Rises Above All," in Seattle *Post Intelligencer*, August 12, 1894.

31 Manuscript notes of Archibald Alston, donated to Mount Rainier National Park by his wife, September 15, 1937.

32 Interview with Mrs. Maude Shaffer (*nee* Longmire) at Tacoma, May 3, 1955.

33 E. C. Smith, Letter to Preston P. Macy, December 29, 1951.

CHAPTER V
pp. 97-118

1 Ernest C. Smith, "An Ascent of Mt. Rainier," in *The Christian Register* (October 3, 1889), 4-5.

2 Mount Rainier had definitely been climbed five times, and probably six, prior to 1889, not including the three ascents which appear to have reached the dome just short of the summit.

3 They were at "Camp Muir," which Smith has confused with the camp of the 1883 party; Camp Ewing was south of the prominent rocky knob on the east side of Camp Muir.

4 Since the blocky cliff was named because of its resemblance to the guardian of the Mediterranean Sea, the spelling should be Gibraltar.

5 Perhaps he drew some consolation from Isaiah 28:20—"For the bed is too short to stretch oneself on it, and the covering too narrow to wrap oneself in it."

6 Now Camp Misery. The name Camp of the Stars was later applied to other sites nearby.

7 August 13, 1889.

8 This was the first party to ascend Mount Rainier with the intention of remaining overnight on the summit, so blankets were carried up. It was also the first overnight stay in the larger crater.

9 Ernest C. Smith, Letter to the author, December 1, 1954.

10 Charles H. Gove, "Night on the summit of Mount Rainier," in *The Mountains of Oregon* (Portland, Oregon, 1890), 43-51. Referred to hereafter without citation.

11 They were attempting to reach Camp Ewing, which they thought was at 11,000 feet, and did not know they were already above it. Their barometer showed 10,500 feet at Camp Muir.

12 "Gove Shot to Kill," in Portland *Oregonian*, March 26, 1893.

13 Christine Louise Van Trump, "Miss Van Trump's Account," in Tacoma *Every Sunday*, November 19, 1892. This is the last issue of *Every Sunday*.

14 Enumclaw *Evergreen*, January 15, 1892. No file of that newspaper has been located.

15 The date of the ascent is given as July 13, 1890. See "Mount Tahoma . . . Historical Sketch of All Successful Climbers to the Above Date," in *Tacomian*, January 14, 1893. The same unidentified source says he remained on top for twenty minutes.

16 According to the topographic map, the east crater is 300 by 400 yards. See *Mount Rainier National Park*, Washington, 1934 ed., 1:62,500 scale, United States Geological Survey.

17 "Mount Tahoma . . . Oscar Brown's Ascent in 1890," in *Tacomian*, March 25, 1893.

18 Fay Fuller, "A Trip to the Summit," in Tacoma *Every Sunday*, August 23, 1890. This account is used frequently hereafter without citation.

19 Amsden took the first photographs made on Mount Baker.

20 "She Reached the Top," in Tacoma *Ledger*, August 17, 1890.

21 Ernest C. Smith, "A Trip to Mount Rainier," in *Appalachia*, VII (March, 1894), 185-205 (read by E. C. Smith before the Appalachian Society, April 11, 1893). This account is used frequently hereafter without citation.

22 They left their names in a sardine can, a brandy flask, and a tin cup.

23 Clara Berwick Colby, "Editorial," in *Women's Tribune*, September 13, 1890.

24 "R. R. Parrish is a Suicide," in Portland *Oregonian*, March 12, 1924.

25 Ernest C. Smith, Letter to the author, December 1, 1954.

26 He is speaking of Reverend E. C. Smith and W. O. Amsden, a Seattle photographer.

27 Near present Camp Comfort at an elevation of 12,000 feet. See "Mount Tahoma," in *Tacomian*, April 1, 1893.

28 Their rations were given as hardtack, ginger snaps, and brandy!

29 Yes. Summit guide Jim Whittaker rediscovered the inscription August 13, 1954. The rock is just below the crater rim on the inside at a point about fifty feet east of the low notch near "Register Rock." On the west face, very neatly done, is the following:

> S. C. Hitchcock
> A. F. Knight
> Will Hitchcock
> Van Watson
> '90

30 Arthur F. Knight, Letter to C. Frank Brockman, August 29, 1933. Copy in the Tacoma Public Library.

31 C. T. Conover, "Mount Rainier and the Facts of History," in *The Great Myth—'Mount Tacoma'* (Olympia, Washington, 1924), 20.

32 *Loc. cit.*

CHAPTER VI
pp. 119-48

1 "Mount Rainier," Vol. I, No. 15, p. 40.

2 "First Flag on Mt. Tacoma," in Tacoma *Union*, June 9, 1895.

3 A. G. Rogers, "A. G. Rogers Account of the 1891 Party," in *Tacomian*, March 25, 1893.

4 A. G. Rogers, "On the Way," in Enumclaw *Evergreen*, June 26, 1891.

5 A. G. Rogers, "Oscar Brown Expedition," in Enumclaw *Evergreen*, July 17, 1891.

227

6 The rusted mattock head brought down from the summit several years ago and given to the Tacoma Mountaineers for their clubroom, may be part of the "pick" used.

7 "Viewed the Volcanoes," in Seattle *Post Intelligencer*, August 5, 1891.

8 Other members of this party climbed to Gibraltar Rock, according to E. Twitmeyer, whose interesting account was published in the *Sharpsville Advertiser* (Penna.) and later reprinted by the *Tacomian*.

9 "Special to the Ledger," in Tacoma *Ledger*, August 4, 1891.

10 In an interview September 28, 1954, I asked Len how many times he climbed Mount Rainier. With a smile he said: "I always say 150, but it probably wasn't that many times."

11 August 23, 1890, "A Trip to the Summit." She wrote: "Year by year it is becoming more difficult along the ledge and those who have made the trip more than once think that soon it will be impossible to go by the old route and that a new way must be discovered. Mr. Van Trump and others are examining the southwest slope and think it may be the coming route."

12 John C. Rathbun, *History of Thurston County*, Washington (Olympia, Washington, 1895) , 95.

13 "The Grand Old Mount," in Tacoma *Ledger*, August 22, 1892.

14 Mrs. George E. Blankenship, *Early History of Thurston County, Washington* (Olympia, Washington, 1914) , 319.

15 P. B. Van Trump, "The North Peak—Van Trump's First Attempt to reach it, in August, 1891," in the *Tacomian*, February 25, 1893. This account is used frequently hereafter without citation.

16 Van Trump's story of the goats paralleling their ascent to the extremity of the ridge which terminates below the Sunset Amphitheater indicates the Tahoma Glacier may have been reached lower than they thought—at an elevation of 10,000 feet rather than 12,000.

17 Rev. J. Warner Fobes, George James, and Richard O. Wells reached the summit August 20, 1884.

18 "What Was Found," in Olympia *Tribune*, August 26, 1891.

19 "Climbing Mount Rainier," in Tacoma *Washington Standard*, September 18, 1891 (reprinted from the *Centralia News*) .

20 P. B. Van Trump, in "On the Mountain Top," in Tacoma *Ledger*, August 31, 1891.

21 "What Was Found," in Olympia *Tribune*, August 26, 1891.

22 "Those Mountaineers," in Tacoma *Ledger*, August 27, 1891.

23 P. B. Van Trump, "On the Mountain Top," in Tacoma *Ledger*, August 31, 1891.

24 The following item appeared in the *Morning Olympian*, September 4, 1891:
Rainier Riley Roasted
The health officer accused in the Council of
neglecting his duty.
If Dr. Riley does not attend to his duties as
health officer he should be removed.

25 Baker's store was on Little Mashel River a mile and a half beyond Eatonville.

26 Tahoma Glacier.

27 Their "west" peak is Point Success, which is actually southwest from the main summit (called the "east" peak here) .

228

28 Columbia Crest.

29 Frank Lowe, "Mount Tahoma," in *The Tacomian*, March 18, 1893 (reprinted from the Orting *Oracle*, which is still published though there is no known file of issues for 1891).

30 "An Easy Trail May Be made Over Gibraltar," in *The Tacomian*, April 1, 1893.

31 They were A. B. McAlpin, C. H. Sholes, W. H. Grant, Geo. A. Steel, N. B. Malleis, and W. H. Steel (publisher of the mountaineering magazine, *Steel Points*, and the memorable book *Mountains of Oregon*). See "Upon Mt. Rainier," in Portland *Oregonian*, September 22, 1891.

32 W. G. Steel, "A Portland Party in 1891 Tries to Reach the Top," in *Tacomian*, December 24, 1892. Manuscript records of the Oregon Alpine Club are in the Oregon Historical Society.

33 "Died a Natural Death," in Portland *Oregonian*, December 12, 1891.

34 Township 15 North, Range 6 East, completed September 27, 1891; and Township 15 North, Range 5 East, completed October 15, 1891; both referenced to the Willamette Meridian.

35 "To the Mines," in Tacoma *Every Sunday*, March 12, 1892. Brown's Junction would later be known as Elbe. "From Mine to Ranch," in Tacoma *Ledger*, July 1, 1892.

36 P. B. Van Trump, "On the Mountain Top," in Tacoma *Ledger*, August 31, 1891.

37 P. B. Van Trump, "Names of the Great Mountain Discussed by Van Trump," in *Tacomian*, December 24, 1892.

38 United States Patent No. 19654, dated February 6, 1892, at 1 p.m., in Book 89, Deeds, Page 99. Fee No. 67302.

39 "Mountain Murmurs," in Tacoma *Every Sunday*, July 2, 1892.

40 *Ibid.*, July 23, 1892.

41 George L. Dickson, "George L. Dickson's Narrative," in *Tacomian*, January 7, 1893.

42 "Climbing Mt. Tacoma," in Tacoma *Ledger*, August 3, 1892.

43 Len Longmire had been to Paradise Park to pick huckleberries and was returning to the springs with his packhorse well laden with fruit when yellow jackets made a sally upon them as they were descending the switchbacks; the horse bucked and the berries returned to nature. In an attempt to eliminate the little nuisances, Len set fire to their nest, but conditions were just right for a rapid spread, and frantic efforts to extinguish the flames were of no avail. Hundreds of acres of beautiful forest went up in smoke.

44 The Ingraham-Smith party was on the summit August 13-14, 1889.

45 "Lots of Mountain Climbers," in Tacoma *Ledger*, July 22, 1892.

46 August 9, 1892, "Tacoma's North Peak."

47 "Mountain Murmurs," in Tacoma *Every Sunday*, August 13, 1892.

48 "Mountain Murmurs," in Tacoma *Every Sunday*, August 20, 1892. Like Dr. Riley's other published accounts, this one is ridiculously muddled and less truth than fancy.

49 "Return of Dr. Riley," in Tacoma *Ledger*, August 12, 1892.

50 P. B. Van Trump, "Conquering the Giant," in Tacoma *Ledger*, September 12, 1892.

51 "Tourist" (to ed.), "False Stories Denied," in Tacoma *Ledger*, August 3, 1892.

52 Cora M. Gordon, "Camp of the Clouds—Paradise Valley," an unpublished manuscript, dated September 24, 1892, now in the Washington Historical Society Library.

53 "Home from the Mountain," in Tacoma *Ledger*, September 1, 1892.

54 "The Quickest Trip on Record From this City to Mount Tacoma," in Tacoma *Ledger*, July 13, 1892; "After Mountain Scenery," in Tacoma *Ledger*, July 10, 1892; "Back from Mount Tacoma," in Tacoma *Ledger*, July 13, 1892.

55 Interview with Mrs. Maude Shaffer (*nee* Longmire), May 3, 1955, at Tacoma, Washington.

56 Interview with Leonard Longmire, October 12, 1954, at Yelm, Washington.

57 "All Night in a Gale," in Tacoma *Ledger*, August 28, 1892; "Mountain Murmurs," in Tacoma *Every Sunday*, September 3, 1892.

58 "Mountain Murmurs," in Tacoma *Every Sunday*, September 3, 1892.

59 "Mountain Murmurs," in Tacoma *Every Sunday*, July 23, 1892; "Lots of Mountain Climbers," in Tacoma *Ledger*, July 22, 1892.

60 Details of this ascent have been drawn from "Conquering the Giant," in Tacoma *Ledger*, September 12, 1892; "Like a flash, an Oaklander's Awful slide down Mt. Rainier," in Oakland, California, *Enquirer*, September 23, 1892; and P. B. Van Trump, "Mount Tahoma," in *Sierra Club Bulletin*, I (May, 1894), 109-32.

61 "Ten Miles of Road," in Tacoma *Ledger*, July 4, 1892.

62 Probably the meadow at the foot of Pyramid Peak.

63 September 12, 1892, "Conquering the Giant."

64 Van Trump was referring to an editorial, "Reserve the Mountain Tops," in Tacoma *Ledger*, August 30, 1892. In it, attention was called to Oregon's efforts to have an area including Mount Hood set aside as a "national reserve." The editor suggested, "It will be possible to have the top of Mount Tacoma reserved this way, unless Seattle may happen to want it for something, and a movement ought to be started to that end."

65 Interview with Len Longmire, September 28, 1954; confirmed by Mrs. Sue Hall (*nee* Longmire), May 24, 1955; see also the interview with Len Longmire in "An Old Timer Recalls Some Interesting Happenings of the Past," *Mount Rainier National Park Nature Notes*, Vol. XI, No. 7 (September, 1933), p. 5.

66 Judge Calkins, "Save Paradise Valley," quoted in Tacoma *Ledger*, September 2, 1892.

67 Sluiskin Mountains. Note how early this name appeared.

68 "A National Park," in Tacoma *Ledger*, September 3, 1892.

69 "Would Lead All Parks," in Tacoma *Ledger*, September 3, 1892.

70 "Schwatka and Mount Tacoma," in Tacoma *Ledger*, September 14, 1892.

71 "Wants Mt. Tacoma Saved," in Tacoma *Ledger*, September 4, 1892.

72 "Lost in a Snow Storm," in Tacoma *Ledger*, September 28, 1892.

CHAPTER VII

pp. 149-85

1 January 15, 1893. Actually, F. S. Emmons published the first map in *The Nation*, No. 595 (November 23, 1876), 312-13. Plummer's map was republished in *Harper's Weekly* during 1897 (p. 675).

2 "The Top of the Mountain," in Tacoma *Ledger,* January 18, 1893.

3 Fred Plummer, "The Top of Tacoma," in Tacoma *Ledger,* January 20, 1893. Plummer has his facts mixed; James Longmire did not climb with Stevens and Van Trump in 1870, and Kautz made his attempt July 16, 1857.

4 "Gen. Kautz Home from Europe," in Tacoma *Ledger,* January 16, 1893.

5 "The Mountain Mapped," in *Tacomian,* January 21, 1893.

6 "First on Tacoma's Top," in Tacoma *Ledger,* January 28, 1893.

7 "Fred Plummer's Response," in Tacoma *Ledger,* January 29, 1893.

8 "He Reached the Top," in Tacoma *Ledger,* January 31, 1893.

9 P. B. Van Trump, "Van Trump Answers Plummer," in Tacoma *Ledger,* February 2, 1893.

10 F. G. Plummer, "A Card from Mr. Plummer," in *Tacomian,* February 11, 1893.

11 "Mr. Ingraham's Letter," in *Tacomian,* February 4, 1893.

12 P. B. Van Trump, "Nameless Forever More," in *Tacomian,* February 11, 1893.

13 J. T. Forrest, "A Guide in Mountain Climbing," in Tacoma *Ledger,* January 27, 1893. The other members of this party were L. A. Davis, High Barnes, Judge Pearson, and Caspar Wrensh.

14 "Mount Tacoma Park," in Tacoma *Ledger,* January 27, 1893.

15 "We'll Have the Park," in Tacoma *Ledger,* January 26, 1893.

16 "Mount Tacoma Park," in Tacoma *Ledger,* January 27, 1893.

17 "Mount Tacoma Park," in Tacoma *Ledger,* February 23, 1893.

18 James D. Richardson, compiler, *Messages and Papers of the Presidents,* IX (Washington, 1903), 779.
The action was taken under section 24 of the Act of Congress approved March 3, 1891, entitled "An Act to repeal timber-culture laws, and for other purposes," which reads, in part: "That the President of the United States may from time to time set apart and reserve . . . [timbered lands] as public reservations; and the President shall by public proclamation declare the establishment of such reservations and the limits thereof."

19 "Noble Erases Rainier," in Tacoma *Ledger,* February 12, 1893.

20 The boundaries of the Pacific Forest Reserve were described as follows: "Beginning at the Southwest corner of township thirteen north, range fifteen east, of the Willamette base and Meridian; thence northerly along the surveyed and unsurveyed range line between ranges fourteen and fifteen east, subject to the proper easterly or westerly offset on the fourth standard parallel north, to the point for the northeast corner of township eighteen north, range fourteen east; thence westerly along the unsurveyed township line between townships eighteen and nineteen north, to the southeast corner of township nineteen, range seven east; thence southerly along the unsurveyed range line between ranges seven and eight east, subject to proper easterly or westerly offsets on the township line between townships seventeen and eighteen north, and the fourth standard parallel north, to the point for the southwest corner of township thirteen north, range eight east; thence easterly along the unsurveyed township line between townships twelve and thirteen north, to the southwest corner of township thirteen north, range fifteen east, the place of beginning . . ." From, "A Bill to Set Apart Certain Lands, Now Known as Pacific Forest Reserve, as a Public Park, to Be Known as the Washington National Park" (Senate Bill No. 1250, 53rd Congress).

21 A. V. Kautz, "He Reached the Top," in Tacoma *Ledger,* January 31, 1893.

231

22 "Two Alpine Clubs Formed," in Tacoma *Ledger,* February 26, 1893.

23 James Van Marter, "Want a Climber's Club?", in Tacoma *Ledger,* February 5, 1893.

24 "Mountain Climbers Organize," in Tacoma *Ledger,* February 10, 1893.

25 "Washington Alpine Club," in Tacoma *Ledger,* February 12, 1893.

26 "Two Alpine Clubs Formed," in Tacoma *Ledger,* February 26, 1893.

27 "First Woman to Climb Mountain Visits City," in Tacoma *News-Tribune,* August 17, 1950.

28 "Out on Business Collecting Fees," in San Francisco *Examiner,* December 5, 1897.

29 "Will Climb Mountains," in Tacoma *Ledger,* June 30, 1893.

30 On September 18, 1927, an organization which formed at Seattle in 1916 as the "Co-Operative Campers of the Pacific Northwest," adopted the name Washington Alpine Club. It may be only a coincidence that one of the founders was Edward Sturgis Ingraham, but I would rather believe he revived the old name out of sentimental attachment for a bygone day.

31 "Would Lead All Parks," in Tacoma *Ledger,* September 3, 1892.

32 "A Mountain Hotel," in Tacoma *Ledger,* February 10, 1893.

33 "Men of the Mountain," in Tacoma *Ledger,* April 4, 1893.

34 "Tacoma's Opportunity," in Tacoma *Ledger,* April 11, 1893.

35 "On to Mount Tacoma," in Tacoma *Ledger,* April 13, 1893.

36 "The Road is Assured," in Tacoma *Ledger,* April 19, 1893.

37 "The Mountain Wagon Road," in *Tacomian,* July 1, 1893.

38 P. B. Van Trump, "The Road to Mount Tacoma," in Tacoma *Ledger,* April 23, 1893.

39 Maj. W. H. Grattan, "The Road to Mount Tacoma," in Tacoma *Ledger,* April 25, 1893.

40 "Roads to the Mountain," in Tacoma *Ledger,* May 26, 1893.

41 "The Mount Tacoma Road," in Tacoma *Ledger,* June 7, 1893.

42 "James Longmire Writes . . ." in *Tacomian,* June 24, 1893; and "The Mountain Wagon Road," in *Tacomian,* July 1, 1893.
Letter of Alfred Lovell to his sister, Maria, August 28, 1893, in which he says, "A road has been cut through the forest 12 miles further this year to Longmire Springs, but it is neither graded nor the stumps taken out, so it is almost impossible to ride in a wagon, although a few have been over it." Published as "Mount Rainier Was Days Away 60 Years Ago," in Tacoma *News-Tribune,* December 20, 1953.

43 H. G. Rowland, "Mountain Trip for Hardy Only in '93," in Tacoma *News-Tribune,* February 13, 1916.

44 "A Photographic Expedition," in Tacoma *Ledger,* August 4, 1893. "They Photographed the Mountain," in Tacoma *Ledger,* August 23, 1893.

45 "New Route to the Summit," in Tacoma *Ledger,* September 6, 1893.

46 "Will Climb Mount Tacoma," in Tacoma *Ledger,* September 3, 1893. George W. Driver appears to have been involved in the Mashel River water scandal which was shaping up at that time.

47 Addie F. Barlow, "Trips to the Mountain," in Tacoma *Ledger,* September 18, 1893.

48 "New Route to Summit," in Tacoma *Ledger*, September 6, 1893. Arthur F. Knight, Letter to C. Frank Brockman, August 29, 1933.

49 "Returned from the Mountain," in Tacoma *Ledger*, August 12, 1893.

50 "Surveying Party Off for Mount Tacoma," in Tacoma *Ledger*, August 23, 1893.

51 *Polk's Tacoma City Directory*, 1893-94, 345.

52 "Slept in the Crater," in Tacoma *Ledger*, October 2, 1893.

53 "Along Paradise River," in Tacoma *Ledger*, September 4, 1893.

54 Mrs. George E. Blankenship, *Early History of Thurston County, Washington* (Olympia, Washington, 1914), 119.

55 "Along Paradise River," in Tacoma *Ledger*, September 4, 1893.

56 Garrett P. Serviss, "Grandest American Mountain," in Tacoma *Ledger*, May 3, 1893.
Henry T. Fink, "Our Grandest Mountain," in Tacoma *Ledger*, June 2, 1893.

57 "Evergreen State Souvenir," in Tacoma *Ledger*, July 9, 1893.

58 "It Will be Called Tacoma," in Tacoma *Ledger*, July 7, 1893.

59 "A Map Raised a Row," in Tacoma *Ledger*, July 3, 1893.

60 "Guide Book to Mount Tacoma," in Tacoma *Ledger*, August 9, 1893. Fred G. Plummer, *Illustrated Guide Book to Mount Tacoma* (Tacoma, Washington, 1893), 12 pp. and map.

61 The present Mountain Highway probably took its name from that earlier Mountain Road, which it replaced, though the routes were different until Elbe was reached.

62 "How High is Mount Tacoma?" in Tacoma *Ledger*, July 6, 1893.

63 "New Route to Summit," in Tacoma *Ledger*, September 6, 1893.

64 "Mt. Tacoma Names," in Tacoma *Ledger*, December 12, 1893.

65 "Glaciers of Mount Tacoma," in Tacoma *Ledger*, July 1, 1893. "Views of Mt. Tacoma's Glaciers," in Tacoma *Ledger*, August 19, 1893. "Off for Mount Tacoma," in Tacoma *Ledger*, September 4, 1893.

66 "Mt. Tacoma Vandals," in Tacoma *Ledger*, September 3, 1893.

67 "Vandalism on the Mountain" (Ed.), in Tacoma *Ledger*, September 4, 1893.

68 "Protection for Paradise Valley," in Tacoma *Ledger*, September 8, 1893.

69 "Vandalism in Paradise Park," in Tacoma *Ledger*, September 9, 1893.

70 "Visitors to the Mountain," in Tacoma *Ledger*, September 13, 1893.

71 "Mount Tacoma Vandals," in Tacoma *Ledger*, September 14, 1893.

72 "Files on the Mashel," in Tacoma *Ledger*, September 14, 1893.

73 53rd Cong., 2nd Sess., *Senate Misc. Doc. No. 247*, introduced July 16, 1894, by Senator Watson C. Squire (Washington).

74 S. 1250, 53rd Congress (See *Congressional Record*, December 12, 1893, p. 154).

75 H. R. 4989, 53rd Congress.

76 Mr. Edmund B. Rogers, Letter to the author, November 19, 1957; and, Commissioner of the General Land Office, Letter to the Secretary of Interior, February 19, 1894.

77 Bailey Willis, *A Yanqui in Patagonia*, 23-24.

78 He was one of the original signers of the articles of incorporation of the Sierra Club, June 4, 1892.

79 "The sad news of the accidental death of George Bayley in San Francisco yesterday . . .," in Oakland *Tribune,* May 1, 1894.

80 E. S. Ingraham, "It Rises Above All," in Seattle *Post Intelligencer,* August 12, 1894, used hereafter without citation.

81 E. S. Ingraham, "From the Ice Caves," in Seattle *Post Intelligencer,* July 25, 1894, used hereafter without citation.

82 Mount St. Elias, being an Alaskan peak, could hardly be considered within the United States; if he really meant North America, Mounts McKinley and Logan would have to be included.

83 "Great Glacier is a Monument to Seattle Man," in Seattle *Times,* July 19, 1925.

84 Miss Helen Holmes was then only fifteen years of age and the youngest of the five women who had reached the top.

85 The name Mazama was taken from a former scientific name applied to the mountain goat (*Oreamnos montanus,* merriam) . *See* Martin W. Gorman, "The word Mazama," in *Mazama,* II (October, 1900) , 46-50. The club was incorporated under the laws of Oregon on March 16, 1899.

86 Olin D. Wheeler, "Mount Rainier, Its Ascent by a Northern Pacific Party," in *Wonderland* (1895) , 53, used frequently hereafter without citation.

87 Wheeler claims that more than 700 persons visited Paradise Park in 1894.

88 The Sarvent map was also published in the *Eighteenth Annual Report of the United States Geological Survey* (1897) , and in *Mazama,* II (October, 1900) , opposite p. 30.

89 S. 2204, 53rd Congress, introduced by Senator Squire (Washington) under the title of the previous bills.

90 United States 53rd Congress, 2nd Session, *Senate Misc. Doc. No. 247;* reprinted in the *Eighteenth Annual Report of the United States Geological Survey, 1896-97,* and in Meany's *Mount Rainier . . .* (1916) , 287-96. (See *Congressional Record,* July 26, 1894, pp. 7877-78.)

91 *Congressional Record,* July 26, 1894, pp. 7877-78.

92 *Congressional Record,* August 9, 1894, p. 8361.

93 *Congressional Record,* December 22, 1894, p. 551.

94 "His Crown is Broken," in Seattle *Post Intelligencer,* November 22, 1894.

95 "Tried Winter Climb, 1894," in Seattle *Times,* February 6, 1922.

CHAPTER VIII

pp. 186-200

1 See *Congressional Record,* January 9, 1895, p. 764; January 19, 1895, p. 1132, and January 24, 1895, p. 1276. 53rd Congress, 2nd Session, *Senate Mis Doc. No. 95.*

2 Information from a note with an 1895 photograph of Comstock's "hotel" (a copy is in the photograph file of Mount Rainier National Park) .

3 E. S. Ingraham, *The Pacific Forest Reserve and Mt. Rainier* (Seattle, 1895) .

4 S. 164, 54th Congress, introduced December 3, 1895, by Senator Squire; and H. R. 327, 54th Congress, introduced December 6, 1895, by Representative Doolittle.

5 H. R. 4058, 54th Congress, introduced January 15, 1896, by Representative Doolittle.

6 See *Congressional Record,* June 10, 1896, pp. 6406-407 (House) .

7 *Ibid.*

8 See *Congressional Record,* March 3, 1897, pp. 2717-18 (Senate) .

9 James D. Richardson, *Messages and Papers of the Presidents,* 1789-1902, IX (Washington, 1903) , 777-79.

10 Olof Bull, "Veteran Violinist Recalls Ascent of Mountain—Diary Kept by Olof Bull contains Records of Climbs in Early Days," in Tacoma *News-Tribune,* August 7, 1920. As mentioned in Chapter V, I do not believe Oscar Brown made a solo climb in 1890.

11 Mrs. S. H. Rathbun *(nee* Hope Willis) , Letter to her sister, Mrs. Donald F. Smith, July 23, 1954.

12 Bailey Willis, Letter to his mother, July 23, 1896.

13 For Russell's remarkably similar account of this near accident, see "Glaciers of Mount Rainier," in *Eighteenth Annual Report of the United States Geological Survey* (1897) , 371.

14 Bailey Willis, "Explorations in the Early Eighties," in *Mountaineer,* VIII (December, 1915), 46-48.

15 Meany, *Mount Rainier,* p. 50.

16 Bailey Willis, Letter to his mother, August 2, 1896.

17 Bailey Willis died February 19, 1949.

18 Edward T. Parsons, "Rainier," in *Mazama,* II (October, 1900), 33.

19 E. H. M'Allister, "Prof. Edgar McClure," in *Mazama,* II:41-45.

20 Edward T. Parsons, "Rainier," in *Mazama,* II (October, 1900), 33.

21 Fred G. Plummer, "Mt. Rainier Forest Reserve, Washington," in *Twenty-First Annual Report of the United States Geological Survey,* Part V (1899-1900), 94.

22 S. 2552, 55th Congress, introduced December 7, 1897, by Senator Wilson.

23 H. R. 9146, 55th Congress.

24 Ernest C. Smith, Letter to Superintendent Preston P. Macy, Mount Rainier National Park, December 29, 1951; Letter to the author, December 1, 1954.

25 John P. Hartman, "Creation of Mount Rainier National Park," an address delivered at the 37th Annual Convention of Washington Good Roads Association, Olympia, Washington, September 27 and 28, 1935.

26 Chapter 337, page 993, 31st Statute at Large *(16 U. S. Code Annotated,* p. 36). The boundaries of the "Public Park to be Known as 'Mount Ranier [*sic*] National Park'," approved March 2, 1899, were described as follows in the organic act: "Beginning at a point three miles east of the northeast corner of township numbered seventeen north, of range six east of the Willamette Meridian; thence south through the central parts of townships numbered seventeen, sixteen, and fifteen north of range seven east of the Willamette Meridian, eighteen miles more or less, subject to the proper easterly or westerly offsets, to a point three miles east of the northeast corner of township numbered fourteen north, of range six east of the Willamette Meridian; thence east on the township line between townships numbered fourteen and fifteen north eighteen miles more or less to a point three miles west of the northeast corner of township numbered fourteen north, of range ten east of the Willamette Meridian; thence

235

northerly, subject to the proper easterly or westerly offsets, eighteen miles more or less to a point three miles west of the northeast corner of township numbered seventeen north, of range ten east of the Willamette Meridian (but in locating said easterly boundary, wherever the summit of the Cascade Mountains is sharply and well defined, the said line shall follow the said summit, where the said summit line bears west of the easterly line as herein determined) ; thence westerly along the township line between said townships numbered seventeen and eighteen to the place of beginning, the same being a portion of the lands which were reserved from entry or settlement and set aside as a public reservation by proclamation of the President on the twentieth day of February, in the year of our Lord eighteen hundred and ninety-three . . ."

EPILOGUE

pp. 201-206

1 Oscar A. Piper, Report to Eugene Ricksecker, March 17, 1904 (Ricksecker papers, Mount Rainier National Park) .

2 Hazard Stevens, "Changes in Mt. Tak-Ho-Ma," in Mazama, II (December, 1905), 202. He makes similar comments in "Changes," in Mountaineer, VIII (December, 1915) , 45.

3 See also Emmons, "The Volcanoes of the Pacific Coast of the United States."

4 Ethan Allen, Report of the Superintendent of the Mount Rainier National Park to the Secretary of the Interior, 1914 (Washington, 1914) , 7.

5 D. L. Reaburn, Report to the Secretary of the Interior by the Supervisor of the Mount Rainier National Park, 1915 (Washington, 1915), 7.

6 The idea was first advanced in "We'll Have a Park," in Tacoma Ledger, January 27, 1893.

7 Interview with Mrs. Susan Hall, nee Longmire, May 24, 1955.

8 Tacoma Tribune, April 18, 1915.

9 Edmond S. Meany, "Memorial Seat at Sluiskin Falls," in Mountaineer, XIV (December, 1921), 51-55.

10 "Death Comes to P. B. Van Trump," in Tacoma Tribune, December 28, 1916.

11 "Indians Defy Park Hunting Regulations," in Tacoma Ledger, September 10, 1915.

12 Loc. cit.

13 D. L. Reaburn, Report to the Secretary of the Interior by the Supervisor of the Mount Rainier National Park, 1915 (Washington, 1915), 10. The supervisor dismissed the trespasses of that year with the following brief statement: "Hunting is absolutely prohibited in park territory, but the densely wooded nature of park territory makes it impossible to entirely stop the practice."

14 The Yakima treaty (12 Stat. 951, June 9, 1885) is reproduced in the booklet The Yakimas (Yakima, Washington, 1955), 21-25.

15 "Not Sluiskin, says Mountain Guide," in Tacoma Ledger, September 12, 1915.

16 "He's Sluiskin, The Old Chief who Guided Stevens and Van Trump in Historic Climb," in Tacoma Ledger, September 19, 1915.

17 December 30, 1917.

18 As the final draft of the manuscript for this book was being typed, Mrs. Fritz von Briesen (nee Fay Fuller) died at Santa Monica, California, May 28, 1958.

BIBLIOGRAPHY

I. Books, Pamhlets, and U. S. Documents.

Allen, Ethan. *Report of the Superintendent of the Mount Rainier National Park to the Secretary of the Interior*, 1914, Washington, 1914.

Bade, William F., ed. *Steep Trails*, Boston, 1918.

Bancroft, Hubert Howe. *History of Washington, Idaho, and Montana, 1845-1849*, San Francisco, 1890.

Bates, H. W., ed. *Illustrated Travels*, Vol. V, London, n.d.

Blankenship, Mrs. George E. *Early History of Thurston County, Washington*, Olympia, Washington, 1914.

Cornelius, Brother. *Keith, Old Master of California*, New York, c. 1942.

Davenport, Charles B. *Biographical Memoir of George Davidson, 1825-1911* (Vol. XVIII of *Biographical Memoirs*), National Academy of Sciences, 1937.

Davidson, George. Directory for the Pacific Coast of the United States (Appendix 39 to the Report to the Superintendent of the United States Coast Survey, 1858), n.p., n.d.

Gerrish, Theodore. *Life in the World's Wonderland*, Biddeford, Maine, 1887.

Hazard, Joseph T. *Snow Sentinels of the Northwest*, Seattle, 1933.

Heitman, Francis B. *Historical Register and Directory of the United States Army*, Vol. I, Washington, 1903.

Hunt, Herbert. *Tacoma, Its History and Its Builders*, 3 vols., Chicago, 1916.

Ingraham, Edward S. *The Pacific Forest Reserve and Mt. Rainier*, Seattle, 1895.

Meany, Edmond S., ed., *Mount Rainier, A Record of Exploration*, New York, 1916.

Muir, John, ed. *Picturesque California, the Rocky Mountains, and the Pacific Slope*, div. 6, New York, c. 1888.

Mumm, A. L., comp. *The Alpine Club Register*, 1857-63, London, 1923.

Oakland City Directory (California), 1888-91.

Olympia Chamber of Commerce and Thurston County Pioneer and Historical Society. *The Great Myth—"Mount Tacoma,"* Olympia, Washington, 1924.

Plummer, Fred G. *Illustrated Guide Book to Mount Tacoma*, Tacoma, c. 1893.

Plummer, Fred G. "Mt. Rainier Forest Reserve, Washington," in *Twenty-First Annual Report*, Part V, United States Geological Survey (1899-1900).

Polk's Portland City Directory (Oregon), 1890-94.

Polk's Tacoma City Directory (Washington), 1891-95.

Rathbun, John C. *History of Thurston County, Washington*, Olympia, Washington, 1914.

Reaburn, D. L. *Report to the Secretary of the Interior by the Supervisor of the Mount Rainier National Park*, 1915, Washington, 1915.

Rensch, H. E. *Mount Rainier; Its Human History Associations*, Berkeley, California, 1935.

Richardson, James D., comp. *Messages and Papers of the Presidents*, Vol. IX, Washington, 1903.

237

Snowden, Clinton A. *A History of Washington*, 6 vols., New York., 1909.

Splawn, Andrew J. *Ka-mi-akin, The Last Hero of the Yakimas*, Portland, Oregon, c. 1917.

Steel, William G., *The Mountains of Oregon*, Portland, Oregon, 1890.

Stevens, Hazard. *Life of General Isaac I. Stevens*, 2 vols., Boston, 1901.

Tolbert, Caroline C. *History of Mount Rainier National Park*, Seattle, 1933.

United States, Department of the Army. *Medal of Honor*, Washington, 1948.

United States, Geological Survey. *Eighteenth Annual Report*, 1896-97.

United States, 31st Statute at Large (16 *U. S. Code, Annotated*).

United States, 53rd Congress, 2nd Session, *Senate Ex. Doc. No. 247*.

United States, 53rd Congress, 3rd Session, *Senate Misc. Doc. No. 95*.

Vancouver, Capt. George. *A Voyage of Discovery to the North Pacific and 'Round the World,* Vol. II, London, 1801.

Wagner, Laura Virginia. *Through Historic Years with Eliza Ferry Leary*, Seattle, 1934.

Whitfield, William. *History of Snohomish County, Washington*, 2 vols., Chicago, 1926.

Wilkes, Charles. *The United States Exploring Expedition*, Vol. IV, Philadelphia, 1845.

Willis, Bailey. *A Yanqui in Patagonia*, Stanford, California, c. 1947.

Winthrop, Theodore. *The Canoe and the Saddle*. New York, 1862.

Wolfe, Linnie Marsh, ed. *John of the Mountains*, Boston, 1938.

II. Magazine Articles.

Bayley, George B., "Ascent of Mount Tacoma," in *Overland Monthly*, VIII [2nd Ser.] (September, 1886), 266-78.

Brown, Allison L., "Ascent of Mount Rainier by the Ingraham Glacier," in *The Mountaineer, XIII* (November, 1920), 49-50.

Emmons, Samuel F., "Glaciers of Mt. Rainier," in *American Journal of Sciences and Arts*, I [3rd Ser.] (March, 1871), 157-67.

Emmons, Samuel F., "Ascent of Mt. Rainier, with Sketch of Summit," in *The Nation*, No. 595 (November 23, 1876), 312-13.

Emmons, Samuel F., "The Volcanoes of the Pacific Coast of the United States," in *Journal of the American Geographical Society* [1877], IX (1879), 45-65.

Flett, J. B., "My First Trip to the Mountain," in *The Mountainer*, VIII (December, 1915), 52.

Fobes, J. Warner, "To the Summit of Mount Rainier," in (Portland) *West Shore*, XI (September, 1885), 265-69.

Gibbs, George, "Physical Geography of the Northwestern Boundary of the United States," in *Bulletin of the American Geographical Society of New York—1872*, 298-392.

Haines, Aubrey L., "John Muir's Ascent of Mt. Rainier, As Recorded by His Photographer, A. C. Warner," in *The Mountaineer*, L (December, 1956), 38-45.

Haines, Aubrey L., "Three Snohomish Gentlemen Make a Successful Ascent" in *The Mountaineer*, XLVII (December, 1954), 5-7, 75-76.

Himes, George H., "Discovery of Pacific Coast Glaciers," in *Steel Points*, I (July, 1907), 146-48.

Himes, George H., "Very Early Ascents," in *Steel Points*, I (July, 1907), 199-200.

Hitchcock, J. R. W., "The Mount Blanc of Our Switzerland," in *Outing*, V (February, 1885), 323-32.

Hittell, John S., "Somes Notes on the Resources, Population and Industry of Washington Territory," in *The Hesperian*, III (November, 1859), 394-411.

Ingraham, Edward S., "Early Ascents of Mount Rainier," in *The Mountaineer*, II (November, 1909), 38-41.

Ingraham, Edward S., "Then and Now," in *The Mountaineer*, VIII (December, 1915), 50-51.

"Journal of Occurrences at Nisqually House, 1833," in *Washington Historical Quarterly*, VI (July, 1915), 184.

Kautz, August V., "Ascent of Mount Rainier," in *Overland Monthly*, XIV (May, 1876), 393-403.

Kautz, August V., "Ascent of Mount Rainier," in *Oregon Native Son*, I (October, 1899), 328-29.

Kautz, Frances, "Diary of Gen. A. V. Kautz," in *Washington Historian*, I (April, 1900), 115.

Leonard, Richard M., and David R. Brower, "A Climber's Guide to the High Sierra," in *Sierra Club Bulletin*, XXV (February, 1940), 41-63.

Longmire, Ben, "A Pioneer Family," in *The Mountaineer*, VIII (December, 1915), 49.

Longmire, Len, "An 'Old Timer' Recalls Some Interesting Happenings of the Past," in *Mount Rainier National Park Nature Notes*, XI (September, 1933), 4-5.

Longmire, James, "Narrative of James Longmire: A Pioneer of 1853," (as told to Mrs. Lou Palmer in 1892), in *Washington Historical Quarterly*, XIII (January and April, 1932), 47-56, 138-50.

M'Allister, E. H., "Prof. Edgar McClure," in *Mazama*, II (October, 1900), 41-45.

McL. Myers, Harry, "On the Ascents of Mount Rainier," in *The Mountaineer*, XIII (December, 1920), 44-47.

McWhorter, Lucullus V., "Chief Sluiskin's True Narrative," in *Washington Historical Quarterly*, VIII (January, 1917), 96-101.

Meany, Edmond S., "Memorial Seat at Sluiskin Falls," in *The Mountaineer*, XIV (December, 1921), 51-55.

Muir, John, "The Ascent of Mount Rainier," in *Pacific Monthly*, VIII (November, 1902), 197-204.

Muir, John, "The Kings River Valley," in *Sierra Club Bulletin*, XXVI (February, 1941), 1-8.

Parsons, Edward T., "Rainier," in *Mazama*, II (October, 1900), 25-34.

Piper, Charles V., "A Narrow Escape," in *The Mountaineer*, VIII (December, 1915), 52-53.

Scott, Leslie M., "News and Comments," in *Oregon Historical Quarterly*, XIX (December, 1918), 337-38.

Smith, Ernest C., "A Trip to Mount Rainier," in *Appalachia*, VII (March, 1894), 185-205.

Stevens, Hazard, "Changes," in *The Mountaineer*, VIII (December, 1915), 45.

Stevens, Hazard, "The Ascent of Takhoma," in *Atlantic Monthly*, XXXVIII (November, 1876), 513-30.

Stevens, Hazard, "Changes in Mt. Tak-Ho-Ma," in *Mazama*, II (December, 1905), 202.

Stevens, Hazard, in *The Nation*, No. 595 (November 23, 1876), 312.

Van Trump, Philemon B., "Mount Tahoma," in *Sierra Club Bulletin*, I (May, 1894), 109-32.

Van Trump, Philemon B., "Mount Rainier," in *Mazama*, II (October, 1900), 1-18.

Van Trump, Philemon B., "Mount Rainier's Three Historic Native Guides," in *The Mountaineer*, VIII (December, 1915), 45-46.

Wheeler, Olin D., "Mount Rainier. Its Ascent by a Northern Pacific Party," in *Wonderland* (1895), 52-103.

Willis, Bailey, "Explorations in the Early Eighties," in *The Mountaineer*, VIII (December, 1915), 46-48.

III.   Newspaper Articles.

Anonymous articles are listed chronologically under the name of the newspaper in which they appeared.

Enumclaw (Washington) *Evergreen:*
"Oscar Brown's Ascent in 1890." January 15, 1892.

Oakland (California) *Enquirer:*
"Like A Flash, An Oaklander's Awful Slide down Mt. Rainier." September 23, 1892.

Oakland (California) *Tribune:*
"The sad news of the accidental death of George Bayley in San Francisco yesterday . . . ." May 1, 1894.

Olympia (Oregon Territory) *Columbian:*
"Visit to Mt. Ranier." September 18, 1852.

Olympia (Washington Territory) *Columbian:*
"Road Over the Cascades." April 23, 1853.
"Down with your dollars." May 28, 1853.
"Road to Walla Walla." June 18, 1853.
"Report . . . exploring party." July 2, 1853.
"Road to Walla Walla." July 16, 1853.

Olympia (Washington Territory) *Echo:*
"Ascent of Mt. Rainier." September 8, 1870.

Olympian (Washington Territory):
"Rainier Riley Roasted." September 4, 1891.

Olympia (Washington Territory) *Pacific Tribune:*
"Ascent of Mount Rainier." August 29, 1870.
"Ascent of Mount Rainier." September 3, 1870.

Olympia (Washington Territory) *Transcript:*
"Ascent of Mount Rainier." August 6, 1870.
"Ascent of Mount Rainier." September 3, 1870.

Olympia (Washington) *Tribune:*
"What was Found." August 26, 1891.

Portland *Oregonian:*
"Alpine Club Organized." September 15, 1887.
"Upon Mount Rainier." September 22, 1891.
"Died a Natural Death." December 28, 1891.
"Gove Shot to Kill." March 26, 1893.
"Nature Lover is Dead." June 4, 1922.
"R. R. Parrish—Suicide." March 12, 1924.

San Francisco (California) *Examiner:*
"Out on Business Collecting Fees." December 5, 1897.

Seattle (Washington) *Post Intelligencer:*
"Viewed the Volcanoes." August 5, 1891.
"From the Ice Caves." July 25, 1894.
"It Rises Above All." August 12, 1894.
"His Crown is Broken." November 22, 1894.
"A Flag Placed on the Summit." January 6, 1895.
"Death Takes E. S. Ingraham." August 17, 1926.

Seattle (Washington) *Times:*
"Tried Winter Climb, 1894." February 6, 1922.
"Great Glacier is a Monument to Seattle Man." July 19, 1925.
"Maj. E. S. Ingraham . . . Dies." August 17, 1926.
"What-a-Man Named Bailey." July 7, 1935.

Snohomish (Washington Territory) *Eye:*
"Rev. J. W. Fobes recently from Syracuse, N. Y. . . ." September 12, 1883.

240

Steilacoom (Washington Territory) *Puget Sound Herald:*
"Personal—Lt. A. V. Kautz." September 24, 1858.
"Important Territorial Road." July 15, 1859.

Tacoma (Washington) *Every Sunday:*
"Mountain Notes." August 23, 1890.
"A Trip to the Summit." August 30, 1890.
"To the Mines." March 12, 1892.
"Mountain Murmurs." July 2, 1892.
"Mountain Murmurs." July 23, 1892.

Tacoma (Washington) *Ledger:*
"She Reached the Top." August 17, 1890.
"Mountain Climbers." August 19, 1890.
"On Mount Tacoma." February 17, 1891.
"Special to the Ledger." August 4, 1891.
"On the Mountain Top." August 26, 1891.
"Those Mountaineers." August 27, 1891.
"Rear-Admiral Rainier." May 9, 1892.
"Ten Miles of Road." June 4, 1892.
"From Mine and Ranch." July 1, 1892.
"After Mountain Scenery." July 10, 1892.
"Back from Mount Tacoma." July 13, 1892.
"Lots of Mountain Climbers." July 22, 1892.
"Climbing Mt. Tacoma." August 3, 1892.
"Tacoma's North Peak." August 9, 1892.
"Return of Dr. Riley." August 12, 1892.
"The Grand Old Mount." August 22, 1892.
"All Night in a Gale." August 28, 1892.
"Reserve the Mountain Top." August 30, 1892.
"Home from the Mountain." September 1, 1892.
"Save Paradise Valley." September 2, 1892.
"A National Park." September 3, 1892.
"Would Lead All Parks." September 3, 1892.
"Wants Mt. Tacoma Saved." September 4, 1892.
"Schwatka and Mount Tacoma." September 14, 1892.
"Lost in a Snow Storm." September 28, 1892.
"The First Map of the Summit of the Old Mountain Ever Published," January
    15, 1893.
"Gen. Kautz Home from Europe." January 16, 1893.
"Mount Tacoma Park." January 27, 1893.
"We'll Have the Park." January 27, 1893.
"Mountain Climbers Organize." February 10, 1893.
"A Mountain Hotel." February 10, 1893.
"Washington Alpine Club." February 12, 1893.
"Noble Erases Rainier." February 12, 1893.
"Mount Tacoma Park." February 23, 1893.
"Two Alpine Clubs Formed." February 26, 1893.
"Men of the Mountain." April 4, 1893.
"Tacoma's Opportunity." April 11, 1893.
"On to Mount Tacoma." April 13, 1893.
"The Road is Assured." April 19, 1893.
"Roads to the Mountain." May 26, 1893.
"The Mount Tacoma Road." June 7, 1893.
"Will Climb Mountains." June 30, 1893.
"Glaciers of Mount Tacoma." July 1, 1893.
"A Map Raised a Row." July 3, 1893.
"How High is Mount Tacoma?" July 6, 1893.
"It Will be Called Tacoma." July 7, 1893
"Evergreen State Souvenir." July 9, 1893.
"A Photographic Expedition." August 4, 1893.

241

"Guide Book to Mount Rainier." August 9, 1893.
"Returned from the Mountain." August 12, 1893.
"Views of Mt. Tacoma's Glaciers." August 19, 1893.
"They Photographed the Mountain." August 23, 1893.
"Walked to Mount Tacoma." August 23, 1893.
"Surveying Party off for Mount Tacoma." August 23, 1893.
"Returned from the Mountain." August 28, 1893.
"To Hunt in the Mountains." September 3, 1893.
"Will Climb Mount Tacoma." September 3, 1893.
"Mt. Tacoma Vandals." September 3, 1893.
"Vandalism on the Mountain." September 4, 1893.
"Off for Mount Tacoma." September 4, 1893.
"Along Paradise River." September 4, 1893.
"New Route to Summit." September 6, 1893.
"Protection for Paradise Valley." September 8, 1893.
"Paradise Valley Vandalism." September 11, 1893.
"Visitors to the Mountain." September 13, 1893.
"Mount Tacoma Vandals." September 14, 1893.
"Files on the Mashel." September 14, 1893.
"On Two Famous Peaks." September 27, 1893.
"Mt. Tacoma Names." December 12, 1893.
"Indians Defy Park Hunting Regulations." September 10, 1915.
"Not Sluiskin Says Mountain Guide." September 12, 1915.
"He's Sluiskin, The Old Chief Who Guided Stevens and Van Trump in Historic
    Climb." September 19, 1915.
Tacoma (Washington) *News-Tribune:*
"No Mountain Trip This Year . . ." April 8, 1915.
"Death Comes to P. B. Van Trump." December 28, 1916.
"Major Ingraham Called by Death." August 17, 1926.
"First Woman to Climb Mountain Visits . . ." August 17, 1950.
Tacoma (Washington) *Tacomian:*
"Historical Sketch of All Successful Climbers to the Above Date." January 14,
    1893.
"Personal." January 21, 1893.
"The Mountain Mapped." January 21, 1893.
"Oscar Brown's Ascent in 1890." March 25, 1893.
"An Easy Trail May be Made Over Gibraltar." April 1, 1893.
"Miscellaneous Mountain Literature . . ." April 1, 1893.
"James Longmire Writes . . ." June 24, 1893.
"The Mountain Wagon Road." July 1, 1893.
Tacoma (Washington) *Union:*
"Planted on Top Mt. Tacoma." May 26, 1895.
"First Flag on Mt. Tacoma." June 9, 1895.
Tacoma *Washington Standard:*
"Climbing Mount Rainier" September 18, 1891
Washington (D C.) *Evening Star:*
[Bailey Willis Map of 1880] January 13, 1894.

By-line articles are listed alphabetically under name of author.

Barlow, Addie G., "Trips to the Mountain," in Tacoma *Ledger,* September 18, 1893.
Bosworth, W. M., "Slept in the Crater," in Tacoma *Ledger,* October 2, 1893.
Bull, Olof, "Veteran Violinist Recalls Ascent of Mountain—Diary Kept by Olof
    Bull Contains Records of Climbs in Early Days," in Tacoma *News-Tribune,*
    August 7, 1920.
Colby, Clara Berwick, "Editorial," in Tacoma *Women's Tribune,* September 13,
    1890.

Conover, C. T., "Just Cogitating," in Seattle *Times,* July 29, 1956.

Dickson, George L., "George L. Dickson's Narrative," in *Tacomian,* January 7, 1893.

Fink, Henry T., "Our Grandest Mountain," in Tacoma *Ledger,* June 7, 1893.

Forest, J. T., "A Guide in Mountain Climbing," in Tacoma *Ledger,* January 27, 1893.

Frost, A. J., W. D., Vaughn, and William Martin, "Old Settler's Stories," in Tacoma *Ledger,* July 22, 1892, February 12, and March 26, 1893.

Fuller, Fay, "A Trip to the Summit," in Tacoma *Every Sunday,* August 23, 1890.

Gove, Charles H., "Night on the Summit," in Tacoma *Ledger,* September 1, 1889.

Grattan, Maj. W. H., "The Road to Mount Tacoma," in Tacoma *Ledger,* April 25, 1893.

Gregg, Kate L., "The Saga of Lolo-Stik: Part I, William Packwood-Pioneer," in Seattle *Times,* October 14, 1951.

Ingraham, Edward S., "Mount Takhoma," in *Tacomian,* December 17, 1892.

Ingraham, Edward S., "Mr. Ingraham's Letter," in *Tacomian,* February 4, 1893.

Ingraham, Edward S., "From the Ice Caves," in Seattle *Post Intelligencer,* July 25, 1894.

Ingraham, Edward S., "It Rises Above All," in Seattle *Post Intelligencer,* August 12, 1894.

James, George, "Mount Rainier," in Snohomish *Eye,* September 6, 1884.

Kautz, August V., "Expedition to the Summit of Mt. Ranier," in Steilacoom *Washington Republican,* July 24, 1857.

Kautz, August V., "The Death of Leschi," in Tacoma *Ledger,* April 9, 1893.

Knight, Arthur F., "Vandalism in Paradise Park," in Tacoma *Ledger,* September 9, 1893.

Longmire, James, "James Longmire, Pioneer," in Tacoma *Ledger,* August 21, 1892.

Lovell, Alfred, "Mount Rainier Was Days Away 60 Years Ago," in Tacoma *News-Tribune,* December 20, 1953.

Lowe, Frank, "Mount Tahoma—An Orting Party Reached the Summit in August, 1891, via the Succotash Valley," in *Tacomian,* March 18, 1893.

McEvoy, Joseph, "An Old Soldier's Story," in Tacoma *Ledger,* May 28, 1893.

Packwood, William, "Important Territorial Road," in Steilacoom *Puget Sound Herald,* July 15, 1859.

Plummer, Fred G., "The Top of Tacoma," in Tacoma *Ledger,* January 20, 1893.

Plummer, Fred G., "A Card From Mr. Plummer," in *Tacomian,* February 11, 1893.

Plummer, Fred G., "Fred Plummer's Response," in Tacoma *Ledger,* January 28, 1893.

Rogers, A. G., "A. G. Rogers' Account of the 1891 Party," *Tacomian,* March 25, 1893.

Rogers, A. G., "On the Way," in Enumclaw *Evergreen,* June 26, 1891.

Rogers, A. G., "Oscar Brown's Expedition," in Enumclaw *Evergreen,* July 17, 1891.

Rowland, H. G., "Mountain Trip for the Hardy Only in '93," in Tacoma *News-Tribune,* February 13, 1916.

Serviss, Garrett P., "Grandest American Mountain," in Tacoma *Ledger,* May 3, 1893.

Smith, Ernest C., "An Ascent of Mt. Rainier," in *Christian Register,* October 3, 1889.

Stampfler, Joseph, "Climbing Mount Tacoma," in Tacoma *News,* July 12, 1907.

Steel, W. G., "A Portland Party in 1891 Tries to Reach the Top," in *Tacomian,* December 24, 1892.

Tourist, "False Stories Denied," in Tacoma *Ledger,* August 13, 1892.

Van Marter, James, "Want a Climber's Club?" in Tacoma *Ledger*, February 5, 1893.

Van Trump, Christine L., "Miss Van Trump's Account," in Tacoma *Every Sunday*, November 19, 1892.

Van Trump, Philemon B., "Mount Takhoma," in Tacoma *Every Sunday*, November 5, 12, 1892.

Van Trump, Philemon B., "On Top of Tacoma," in Tacoma *Ledger*, August 19, 1888.

Van Trump, Philemon B., "On the Mountain Top," in Tacoma *Ledger*, August 31, 1891.

Van Trump, Philemon B., "Conquering the Giant," in Tacoma *Ledger*, September 12, 1892.

Van Trump, Philemon B., "Names of the Great Mountain Discussed by Van Trump," in *Tacomian*, December 24, 1892.

Van Trump, Philemon B., "The Top of the Mountain," in Tacoma *Ledger*, January 18, 1893.

Van Trump, Philemon B., "First on Tacoma's Top," in Tacoma *Ledger*, January 28, 1893.

Van Trump, Philemon B., "Van Trump Answers Plummer," in Tacoma *Ledger*, February 2, 1893.

Van Trump, Philemon B., "Nameless Forever More," in *Tacomian*, February 11, 1893.

Van Trump, Philemon B., "The Road to Mount Tacoma," in Tacoma *Ledger*, April 23, 1893.

Van Trump, Philemon B., "The North Peak—Van Trump's First Attempt to Reach It, in August, 1891," in *Tacomian*, February 25, 1893.

Van Trump, Philemon B., "A Legend of Mount Tacoma," in Tacoma *Ledger*, July 30, 1893.

Van Trump, Philemon B., [Letter to George B. Bayley, dated August 26, 1888], in Oakland *Tribune*, July 23, 30; August 6, 30, 1939.

IV.   Unpublished materials.

Allen, Edward T., Letter to Superintendent O. A. Tomlinson, Mount Rainier National Park, March 10, 1936.

Alston, Archibald, Manuscript notes donated to Mount Rainier National Park by his wife, September 15, 1937.

Bjarke, Nels, "The Indian Henry Trail," mimeographed by the Fern Hill Historical Society, Tacoma, 1940.

Flint, A. L., Letter to Julius Halverson, August 26, 1938, cited by R. N. McIntyre, in *Short History of Mount Rainier*, Longmire, Washington, 1952.

Fuller, Fay (Mrs. Fritz von Briesen), Interviewed by Messrs. Bill, Potts, and McIntyre, of Mount Rainier National Park, August 17, 1950, at Tacoma.

Gibbs, George, "Report to Capt. George B. McClellan, at Olympia, W. T., March 4, 1854," cited by R. N. McIntyre, in *Short History of Mount Rainier*, Longmire, Washington, 1952.

Gordon, Cora C., "Camp of the Clouds—Paradise Valley," an 8-page typescript dated September 24, 1892, in the collection of the Washington Historical Society, Tacoma.

Hall, Mrs. Sue (*nee* Longmire), Interviewed by the author, May 24, 1955, at Tacoma.

Hartman, John P., "Creation of Mount Rainier National Park," an address delivered at the 37th Annual Convention of the Washington Good Roads Association, Olympia, Washington, September 27 and 28, 1935.

Kautz, August V., Manuscript Journal of Lt. August V. Kautz, 1857-1861, in the National Archives, Washington. D. C.

244

Knight, Arthur F., Letter to C. Frank Brockman, August 29, 1933 (copy in the Tacoma Public Library).

Longmire, Len, Interviewed by the author, September 28, October 12, November 21, 1954.

Mazamas, Portland, Oregon, Scrapbook of the Oregon Alpine Club, Vol. I, No. 15.

Peterman, Mrs. Donald (nee Rachel Allen), Photograph album donated to Mount Rainier National Park.

Piper, Oscar A., Report to Eugene Ricksecker, March 17, 1904 (Ricksecker papers, Mount Rainier National Park).

Preston, John (superintendent of Mount Rainier National Park), Letter to E. C. Smith, November 29, 1943.

Rathbun, Mrs. S. H. (nee Hope Willis), Letter to her sister, Mrs. Donald F. Smith, July 23, 1954.

Rogers, Edmund B., Letter to the author, November 19, 1957.

Shaffer, Mrs. Maude (nee Longmire), Interviewed by the author May 3, 1955, at Tacoma.

Smith, Ernest C., Letter to the author, December 1, 1954.

Smith, Ernest C., Letter to Superintendent Preston P. Macy, Mount Rainier National Park, December 29, 1951.

Tolmie, William Fraser, Manuscript Diary of Dr. William Fraser Tolmie, 1833; photostats from original pages 12-21, from the Provincial Archives, Victoria, B. C.

Whittaker, Jim, Note to the author concerning inscriptions copied from a rock on the crater rim, August 13, 1954.

Willis, Bailey, Letter to his mother, July 23, 1896.

Willis, Bailey, Letter to his mother, August 2, 1896. (Bailey Willis papers, Mount Rainier National Park.)

INDEX

Accidents resulting in deaths, 95, 197; injuries, 48, 50, 132, 138, 143, 165; narrow escapes, 76, 167, 191-92, 197
Ainsworth, F. H., 190-93
Ai-yi (Fish Lake), 16
Allen, Edward J., 13
Allen, Edward T., 122
Allen, Grenville F., 201
Allen, Prof. O. D., 97, 201
Alling, Frank, 114, 132
Alpine plants collected by Tolmie, 7-8
Alps (European), mentioned, 39, 45, 140
Alston, Archibald, Jr., 94-95
Alta Vista, mentioned, 89, 110, 180
Altitude, physiological effects of, 127
American Alpine Journal, x
American Association for the Advancement of Science, 173, 175
American Geographical Society, 57
Amsden, W. O., 109, 111-16, 198-99, 227
Anderson and Borrenger (Tacoma jewelers), 115
Anderson, J. F., 196
Ansley, H. C., 197
Appropriations (park), 188, 199-201
Anti-Chinese riots (Seattle), 81
Anvil Rock, 43, 84, 111
Appalachian Mountain Club, 175, 198
Arnold, Charlie, 162, 164
Ascent of Mount Rainier: considered by Tolmie, 8; attempted in 1852, 10-13; probably made in 1854, 15-18; nearly made in 1857, 21-28; definitely made in 1870, 30-51, 51-57; made in 1883, 60-68; made in 1884, 70-79; made in 1886, 81-83; made in 1888, 84-94; made in 1889, 97-102, 102-105; made in 1890, 106-108, 109-15, 115-17; made in 1891, 120-22, 122-23, 123-29, 129-32; made in 1892, 134-36, 136-37, 137-38, 138, 140-41, 141-45, 146; made in 1893, 165-67; made in 1894, 175-78, 179-81; by Mazama Club in 1897, 194-97
Ashford, Washington, 121, 133, 162-63, 201
Ashford's field, 133
Ashford's hotel, 169
Autier, Michael, 190, 193-94
Automobile, first to Paradise, 203
Avalanches, mentioned, 43, 113; on Willis Wall, 75

Babcock, Ruth M., ix
Backbone Ridge, 53
Bailey, Representative (Texas), 188
Bailey, Robert S.; biographical data, 12; mentioned, 10
Bailey Willis map (1880), 83
Bailey Willis trail, built, 59; mentioned, 72, 81, 93-94, 96, 121, 161, 184, 190
Baker, Lt. _____, 1
Baker, Holland W., 135

Baker's store (Meta), 129, 135-36, 169, 228
Balsley, Silas, 120
Bark town (Grindstone Camp), 72, 223
Barlow, Jessie, 164
Barnes' guidebook, 187
Barometers; aneroid, 36, 65, 73, 75-76, 115, 117, 137, 176-77; mercury, 52-53, 111, 196
Bass, Dan W., 85-86, 90, 93
Bathing, 36, 38, 50, 103, 135, 162, 169
Battles: Second Bull Run, 31; Grande Ronde, 20-21, 39; Mashel River, 20, 36, 220; White River, 20
"Battleship Prow" (Steamboat Prow), 83, 191
Bayley, Miss _____, 141
Bayley, George B.: biographical data, 60; climbing in 1883, 60-68; climbing in 1892, 141-45; injury of, 143-45; death of, 175; mentioned, 12, 69, 71, 77, 79, 86, 92, 127, 131, 138, 141, 164, 174
Beal, Prof. W. J., 186
Bear (black), 11, 74
Bear Prairie; naming of, 29; mentioned, 38, 40-41, 48-49, 63, 139
Bear Prairie Point, 53
Bebb, M. S., 102
"Beehive," 180
Bell, Pvt. _____, 22, 27
Bench Lake, 217
Bender, Ross, x
Benson's Hill, 169
Bergh, _____, 189
Berghaus, Dr. Heinrich, 83, 199
Billings, Charles A., 81
Bismuti, Gene, x
Blaine Glacier, 152
Blankenship, G. C., 29
Bolon, A. J., 18
Booth, Norman O.; death of, 226; mentioned, 85-86, 93
Borrenger, Mr. _____, 116-17
Bosworth, William M., 165
Boyd, R. H., 184
Brass plate (1870 ascent), 47, 127, 202
Bresacker, Eugene, 119
Bridges, Carbon River, 72; Nisqually River, 61
British Alpine Club, 32
British Columbia Provincial Library, x
Bronson, Ira, 178
Brown, Allison L., 81-83
Brown, Prof. J. E., 195
Brown, Oscar: as first park ranger, 202-203; climbing in 1890, 106-108; climbing in 1891, 120-22; mentioned, 123, 131, 133, 166
Brown's Junction (Elbe), 133, 229
Bryce, Prof. James, 69
Buckley, General Manager for NP R.R., 70
Buckley, Washington, 121

246

248